# Estimating the impacts of urban growth on future flood risk
# A comparative study

**Cover:**
3d impression of the city of Dhaka, Bangladesh with the 2004 flood extent projected on the eastern part of the city.

# Estimating the impacts of urban growth on future flood risk
# A comparative study

**DISSERTATION**

Submitted in fulfillment of the requirements of
the Board for Doctorates of Delft University of Technology
and
of the Academic Board of the UNESCO-IHE
Institute for Water Education
for
the Degree of DOCTOR
to be defended in public on
Wednesday, 22 November 2017, at 12:30 hours
in Delft, the Netherlands

by

**Willem Veerbeek**
Master of Science in Architecture, Urbanism and Building Sciences, TU Delft
born in Doetinchem, the Netherlands

This dissertation has been approved by the supervisor:
Prof. dr. C. Zevenbergen

Composition of the doctoral committee:

| | |
|---|---|
| Chairman | Rector Magnificus, TU Delft |
| Chairman | Rector UNESCO-IHE |
| Prof. dr. C. Zevenbergen | UNESCO-IHE / TU Delft, supervisor |

Independent members:

| | |
|---|---|
| Prof. dr. ir. V. I. Meyer | TU Delft |
| Drs. W. Ligtvoet | PBL Netherlands Environmental Assessment Agency |
| Prof. dr. R. Ranasinghe | UNESCO-IHE / University of Twente |
| Prof. dr. F. Klein | TU Delft |
| Prof. dr. ir. M. Kok | TU Delft |
| Prof.dr.ir P. van der Zaag | UNESCO-IHE / TU Delft,  Reserve member |

CRC Press/Balkema is an imprint of the Taylor & Francis Group, an informa business

Published by:
CRC Press/Balkema
Schipholweg 107C, 2316 XC, Leiden, the Netherlands
Pub.NL@taylorandfrancis.com
www.crcpress.com – www.taylorandfrancis.com
ISBN 978-0-8153-5733-9

# Summary

## The rise of megacities

Both in scientific as in popular media, the potential future impacts of climate change are extensively covered. The expected trend changes and amplified extreme weather events change the natural hazard profile in many areas of the world including those where the majority of the world's population resides: in cities. Yet, those same areas are witnessing another transformation with a potentially even more profound impact. Many of the world's urban areas are growing at an unprecedented rate. This has led to the emergence of megacities with populations of 10 million or more. Although megacities already appeared in the 1950s with the growth of the New York and Tokyo metropolitan area, the rise of megacities really took off in the 1980s. Currently, the world hosts about 29 megacities of which the majority are located in Asia. This number is expected to increase significantly in the coming decades; roughly every decade 6 new megacities appear.

Many of those megacities are located along major rivers which in many cases exposes an increasing number of people and assets to floods. This is especially the case in rapidly urbanising river deltas, like for instance the Pearl River Delta. Apart from cities like Guangzhou and Hong Kong, the area was predominantly rural untill the early 1990s. Yet, currently this area hosts a network of cities of which the combined population exceeds 57 million inhabitants. Apart from increasing exposure to riverine flooding, extensive urban growth also leads to rainfall induced flooding of built-up areas. This is especially the case in metropolitan areas, where infill or compact extension of built-up areas change the fundamental drainage characteristics. While the effect of urban growth on flood risk is well understood, extensive comparative studies are few in number. Furthermore, future projections are often limited to statistical extrapolations and lack the spatial attributes which seem essential when assessing flood risk; floods are local phenomena. As a consequence, urban growth projections need to be spatially explicit in order to express the differentiation in land use and land cover (LULC) within and between rapid growing urban agglomerations. This is the main objective of this study, which centres around the question: What is the impact of future urban growth on the development of riverine and pluvial flood risk of the fast growing metropolitan areas and how do these compare?

# Projecting future growth

In this study such an assessment is made for 19 fast growing megacities: 15 in Asia, 2 in Africa, 1 in Latin America and 1 in Europe. The growth scenarios comprise of a future extrapolation of historic spatial development trends, and can therefore be character-ised as a business-as-usual (BAU) scenario for urban growth. In order to identify and extrapolate those spatial trends, an urban growth model has been developed which for a given metropolitan area, attempts to derive the underlying rules that lead up to observed LULC transitions. By using a memetic algorithm-enhanced auto-calibration sequence a series of weights are optimized until the model is able to mimic LULC tran-sitions between two base years. Once the model is able to correctly 'predict the past' (i.e. the 2010 LULC map), future projections are developed based on 5-year intervals. The machine learning-based calibration using local data, the initially generic model be-comes specifically suited to develop a BAU-scenario for a particular case. Consequent-ly, 19 case-specific growth models have been developed. To ensure a consistent and uniform approach, only data sources with global coverage has been used. The horizon has been set at 2060, a 50 year projection period which by far exceeds the 20 year planning horizon encountered in some of the case study areas.

The characteristics of the produced growth projections and subsequent LULC distri-butions differ significantly. This is not simply due to growth differentiation between cities (i.e. different spatial trends) but also due to geographic features that define the suitability to host built-up areas. Some geographic locations are spatially constrained, which 'squeezes' urban development into the sparsely available land. This is for in-stance the case in Seoul, where growth is only possible in the narrow valleys or on the wetlands along the coastal areas in the West. In other cases, space is simply not avail-able anymore and urban growth can only occur by leapfrogging development. This is the case in Mumbai, where the existing peninsula is saturated with high density built-up areas. Apart from growth constraints, specific features in cities also act as attrac-tors. For Beijing this is simply the primary urban centre, which causes an almost con-centric urban development. In other cities, primary infrastructure gives rise to ribbon development along major highways. This can be observed in Ho Chi Minh City, Lahore and Tehran. With the exception of Cairo and Calcutta, urban growth is not directed by the proximity of major rivers or streams. This outcome seems remarkable, since water is often considered a guiding feature for urban development. Over the course of the projection period, the growth rates of all megacities in this study decline over time. Yet metropolitan areas of Dhaka, Ho Chi Minh City and especially Lagos are still projected to double in size in less than 35 years. Other cities, like Istanbul, Mumbai and Seoul, grow at a much lower rate with doubling periods of 60 year or more. Apart from significant differentiation in growth rates the composition of the built-up areas also changes. This is primarily illustrated by the significant densification that is projected for the majority of the megacities in this study. Only a few cities (e.g. Shanghai, Calcutta and Mumbai) show significant levels of urban sprawl over the projection period. Although the analysis shows many communalities, no overall trends have been observed that are representative for all cities. This outcome underlines that growth trends cannot be transferred one to one between cities; every city exhibits unique growth features that can only be evaluated on individual basis.

## Impacts on future flood risk

The estimation of the impact of future urban growth of each city on future flood risk is limited to riverine and pluvial flooding. Even though no spatially explicit growth projections have been used, the future impact of coastal flooding has been already covered in other studies. The assessment of riverine flood risk has been based on data produced by the GLOFRIS model, a global river model from which inundation maps has been produced for flood events associated to return periods between 10 and a 1000 years. Since the level-of-detail between the inundation maps and growth projections differs significantly (about 800m and 30m cells respectively), a sensitivity analysis has been performed to check if this discrepancy is a cause for bias or significant errors. This appeared only in the case of a few cities: Shanghai where floods are characterized by very low inundation depths, as well as Istanbul and Tehran where flood exposure is very limited due to the location of rivers. The LULC-based growth projections are unfit to perform detailed flood impact assessments. The concept of flood risk has therefore been limited to assessing the urban flood extent: the estimated intersection of the projected urban extent for a given future year and the flood extent associated to the respective return periods.

For all cities, the outcomes show a large increase of the urban flood extent as the projected growth developed towards the 2060 horizon. This is especially for Dhaka, Ho Chi Minh City and Lahore, where the urban flood extent is projected to more than triple. When ranking on size of the urban flood extent, the list is dominated by the Guangzhou-Shenzhen metropolitan area followed by Calcutta, Beijing, Shanghai and Delhi with urban flood extents covering several hundred squares of kilometres. Especially for Beijing and Shanghai though, the floods are relatively shallow and the outcomes might therefore be overestimated. To validate and extend some of the outcomes, a more in-depth approach has been taken for the assessment of riverine flooding in Dhaka, where apart from the flood extent also the damages have been estimated using an alternative model, depth-damage curves and extended LULC maps in which built-up areas were subdivided into 10 density classes. The outcomes show that the estimated flood damages in 2050 could be up to 6.8 times higher than those estimated for the reference year 2004. This values is in line with the estimated increase of the urban flood extent, which is projected to increase by a factor 6.4. Of major concern is the disproportionate growth of the urban flood extent observed in some of the growth projections. The growth distribution in these cities is shifted towards rapid develop-ment in the floodplains; growth in 'safer' areas occurs at a more modest rate. This is especially for Dhaka, Ho Chi Minh City and Lahore, and to a lesser extent for Jakarta. A disproportionate growth of the urban flood extent signifies a transition in the flood characteristics; current flood risk not only becomes more widespread, but also covers relatively large areas of the urban agglomeration. Integration of flood risk into urban zoning, planning policies and growth containment plans, seems especially prudent for such cities since urban development might lead to a considerable aggravation of flood conditions. On the other hand, there are also cities where the projected growth pro-ceeded more rapidly outside the floodplains. This is for Shanghai, Seoul and Mexico City. Apparently, these cities have a tendency to grow in a flood sensitive manner. Mit-igation of future flood risk might therefore focus on more traditional flood protection measures. Finally, the impact of urban growth on riverine flooding is compared to the sensitivity to shifting flood frequencies, which might act as a proxy indicator for future climate change induced changes in river discharge. Cities that are more sensitive to these shifts are Lagos, Mexico City and Seoul.

The assessment of pluvial flood risk has been limited to a set of proxy indicators. The requirements for 1d2d coupled hydraulic models that are typically used for such as-sessments, made their use infeasible for the vast areas covered by urban extent in the case studies. As an alternative, drainage conditions have been characterised by a set of indicators that cover macro-, meso- and microscale. By determining if future trend changes in the projected LULC transitions occur, the assumption is that local drain-age characteristics could be evaluated. At macro level, the impacts of urban growth estimating growth induced changes in the impervious surface ratio (ISR) of the urban footprint. To express the impacts of changes in the spatial distribution of urban built-up areas on drainage capacity, the fractal dimension (FD) and open land fraction have been used as mesoscale indicators. These express to what degree built-up areas are fragmented and "perforated", which determines the capacity for water storage which is especially important during peak rainfall events. At a microscale, a quantitative anal-ysis of LULC transitions has been made tracing the actual transitions to built-up areas. Does urban densification for instance primarily transition from suburban areas or are high density clusters built up directly from grassland, barren land or other rural open land? After evaluating these indicators individually, they are summarised and combined in a semi-quantitative manner to create an outlook for each city.

Cities with a rrelatively positive outlook are Ho Chi Minh City, Jakarta and Mumbai. The growth projections for these cities do not lead to vast contiguous high density built-up areas. For Ho Chi Minh City this is particularly a product of the high level of fragmentation of the projected future built-up areas. This ensures sufficient water storage capacity. Also in Jakarta and Mumbai this is the case, although for Mumbai the increasing fragmentation occurs in adjacent areas outside of the peninsula, where the main core of the city is located. Cities that score particularly low in this assessment are Karachi and Istanbul. The projections for both cities show a very compact development, resulting in a rapidly increasing mean ISR, little fragmentation and open land and a disproportionate rate of projected development into high density built-up areas. The Guangzhou-Shenzhen area as well as the metropolitan areas of Manila, Mexico City and Dhaka show the overall largest increase in mean ISR, with an increase of 25% or more. Obviously, the assessment ignores the extensive differences in precipitation associated to the different case studies. Rainfall patterns for Tehran differ for instance dramatically from those for Manila. Yet, as for riverine flooding, the assessment is mostly focussing on relatively changes; so do the drainage characteristics fundamentally deviate from the current trends? For Karachi and Istanbul this indeed seems the case. Also the performance of Dhaka and to a lesser extent, Delhi, Guangzhou-Shenzhen and Tehran degrades disproportionately over the projection period.

While the outcomes show alarming trends for both riverine and pluvial flood risk, a combined assessment cannot be directly derived from the outcomes. Apart from the differences in magnitude and exposure, the different approaches in both assessments prevent combining the outcomes. Nevertheless as a product of urban growth, Dhaka, Guangzhou-Shenzhen and to a lesser extent Lahore show a disproportionate increase of the future susceptibility to floods in both domains. Especially in these cities, urban planning could play an important role in limiting future flood risk.

## Applications beyond flood risk

The application of urban growth scenarios in environmental assessment studies is obviously not limited to flood risk only. To illustrate potential applications, two additional studies have been conducted that merely serve as proof-of-concepts. The first study focusses on the potential increase of pollution loads in streams due to rapid slum development. Also here, the spatial distribution of the projected growth determines where and by which magnitude future pollution loads increase. Using the growth projections of slum areas in Lagos as an example, a number of streams and outlets are facing a disproportionate increase, which could have a devastating effect on the receiving estuaries, that play an important role in the ecosystem and often provide the livelihood for a vast number of people. For the second case, urban growth projections have been used to estimate changes in the urban heat island of Mumbai, which could affect local precipitation patterns. Using a series of recorded rainfall events, a 3d mesoscale atmospheric model in which the altering LULC maps act as one of the drivers, the overall rainfall shifted in most cases to higher intensity levels including the peak levels. Although the outcomes cannot be generalized, partially due to the particular topographic conditions, the study shows how urban growth also affects the hazard component of flood risk; instead of only affecting the exposure and sensitivity to floods, urbanization also intensifies local precipitation. These applications show that the availability of explicit urban growth scenarios can provide a foundation for all types of long term environmental assessments. They can provide a baseline from which the impact of alternative policies can be estimated and serve as an alternative to simply using today's conditions as a point of reference.

## Conclusions

To better facilitate the integration of alternative planning policies, strategies and concrete planning measures (e.g. excluding areas from development), the model still requires further improvements. Also the usability of the model needs to be enhanced to ensure application outside the current research domain. Apart from model improvements, an immediate research priority is to integrate climate change scenarios into the assessment and to compare the estimated impacts to those from the urban growth projections. This finally sheds light on how to compare the two. Especially for riverine flooding, this should be a straightforward procedure once the required flood inundation maps are available.

To increase the impact of this study, a proper forum needs to be found beyond the scientific domain. Although extensive networks exist in which focus on urban climate adaptation, disaster management or sustainable development support is typically restricted to individual cities; comparative studies are not necessarily a priority. Yet, such studies are essential in showing the importance of urban planning in limiting the future impacts of natural hazards and to prioritize efforts towards particular urban agglomerations.

# Samenvatting

## De opkomst van megasteden

Zowel in de wetenschap als in de media wordt regelmatig verslag gedaan van de poten-tiële gevolgen van klimaatverandering. De verwachte trendwijziging en toename van extreem weer wijzigt het kwetsbaarheidsprofiel van veel regio's in de wereld, inclusief die regio's waar het overgrote deel van de wereldbevolking momenteel leeft: in st-eden. In diezelfde regio's vindt momenteel echter ook een andere transformatie plaats met wellicht nog grotere consequenties. Veel verstedelijkte gebieden in de wereld groeien momenteel met een ongekende snelheid. Dit leidde tot de ontwikkeling van megasteden met populaties van 10 miljoen inwoners of meer. Hoewel het fenomeen van megasteden haar intrede maakten in de jaren 50 met de groei van als stedelijke agglomeraties als New York en London, begon de onstuimige groei van megasteden pas echt in de jaren 80. Momenteel zijn er circa 29 megasteden op aarde waarvan het merendeel zich bevindt in Azië. Volgens verwachting groet dit aantal gestaag in de komende decennia; typisch komen er 6 nieuwe megasteden per decennium bij.

Een groot aantal megasteden bevinden langs grote rivieren waardoor in veel gevallen een groeiend aantal mensen en vastgoed wordt blootgesteld aan overstromingen. Dit is vooral het geval in snel verstedelijkte deltagebieden zoals bijvoorbeeld de Pearl River delta. Met uitzondering van steden als Guangzhou en Hong Kong, was deze delta be-gin jaren 90 nog nagenoeg ruraal. Momnteel huisvest diezelfde delta een netwerk van steden met een gezamenlijke populatie van meer dan 57 miljoen inwoners. Behalve een toenemende blootstelling aan overstromingen vanuit de rivieren, leidt stedelijke groei tevens tot een toename van wateroverlast als gevolg van lokale neerslag. Dit is met name het geval in stedelijke gebieden waar verdichting en compacte uitbreidingen de drainagekarakteristieken wezenlijk veranderen. Hoewel er voldoende inzicht is over de effecten van stedelijke groei op overstromingsrisico's, is het aantal vergelijkende studies gering. Daarbij komt dat stedelijke groeiprojecties voor de toekomst vaak bep-erkt blijven tot statistische extrapolaties van groeicijfers waardoor de ruimtelijke di-mensie van stedelijke groei die essentieel zijn voor het bepalen van overstromingsri-sico's worden genegeerd. Overstromingen zijn echter vaak lokale gebeurtenissen met specifiek lokale karakteristieken. Daarom is het noodzakelijk om expliciet ruimtelijke projecties voor stedelijke groei te ontwikkelen om zodoende de differentiatie in land-gebruik binnen en tussen snelgroeiende stedelijke agglomeraties uit te drukken. Dit is de primaire doelstelling van deze studie, die zich richt op de vraag wat de invloed van toekomstige stedelijke groei is op de ontwikkeling van overstromingen vanuit rivieren en vanuit lokale neerslag en hoe die zich verhouden tot elkaar?

## Stedelijke groeiprojecties

In deze studie wordt een vergelijking gemaakt tussen 19 snelgroeiende megasteden: 15 in Azië, twee in Afrika en één in Europa. Het ontwikkelde groeiscenario is gebaseerd op extrapolaties van historische ruimtelijke ontwikkelingstrends en kan daarom worden gekarakteriseerd als een "business-as-usual" (BAU)-scenario voor stedelijke groei. Om ruimtelijke trends te identificeren en te extrapoleren is een stedelijk groeimodel ontwikkeld dat voor een gegeven stedelijke agglomeratie de onderliggende regels probeert af te leiden die ten grondslag liggen aan transities in landgebruik. Kalibratie van het model is gebaseerd op een optimalisatie van een reeks gewichten die bij een correcte parametrisering de geobserveerde landgebruik transities tussen twee gegeven basisjaren zo exact mogelijk trachten na te bootsen. Wanneer het model in staat is de transities uit het verleden met een minimale foutmarge te voorspellen, kan het worden ingezet voor het ontwikkelen van toekomstige projecties op basis van intervallen van vijf jaar. Deze op "machine-learning"-gebaseerde kalibratie, maakt gebruik van lokale data waardoor de groeimodellen worden geoptimaliseerd voor landgebruiktransities voor een specifieke locatie. Dit heeft geresulteerd in de ontwikkeling van 19 afzonderlijke stedelijke groeimodellen. Om consistentie en een uniforme benadering te waarborgen is enkel data gebruikt met wereldwijde dekking. De termijn waarvoor de projecties zijn ontwikkeld bedraagt 50 jaar. Deze termijn is substantieel langer dan de termijn van 20 jaar die veelal wordt gehanteerd.

De resulterende geografische patronen van de stedelijke groeiprojecties, verschillen significant per regio. Dit is niet enkel te wijten aan de onderlinge differentiatie in historische groeipatronen tussen de steden, maar tevens aan de geografische context die de geschiktheid bepaald voor herbergen van stedelijke groei. In sommige locaties wordt de groei bijvoorbeeld beperkt door een beperkte hoeveelheid land dat geschikt is voor urbanisatie. Dit is het geval in Seoul waar groei enkel mogelijk is aan de voet van de steile bergwanden of aan de aan de laaggelegen natte kustzone aan de westkant. In andere gevallen is land voor stedelijke uitbreiding simpelweg niet meer voorhanden, waardoor stedelijke groei slechts kan plaatsvinden op nieuwe locaties die niet verbonden zijn met de bestaande stad. Dit is bijvoorbeeld het geval in Mumbai, waar het bestaande schiereiland vrijwel volledig bestaat uit een stedelijk landschap met zeer hoge dichtheid. Naast groeibeperkingen, kunnen specifiek geografische eigenschappen van steden ook juist fungeren als attractoren van groei. Dit is het geval in Beijing, waar het primaire stedelijke centrum fungeert als een zwaartepunt waaromheen

een bijna volmaakte concentrische ontwikkeling plaatsvindt. In andere steden is het vooral infrastructuur die leidend is voor stedelijke groei. De groei in Ho Chi Minh City, Lahore en Teheran wordt gekenmerkt door een hoge mate van lintbebouwing langs het hoofd wegennetwerk. Met uitzondering van Cairo en Calcutta, wordt de groei in geen van de steden gerealiseerd langs rivieren of secundaire watergangen. Dit lijkt opmerkelijk aangezien water vaak als structurerend element wordt gezien voor stedelijke ontwikkeling. De groeisnelheid neemt voor alle steden substantieel af. Desondanks voorspellen de projecties voor Dhaka, Ho Chi Minh City en Lagos nog steeds een verdubbeling van de grootte binnen een termijn van 35 jaar. Voor andere steden zoals Istanbul, Mumbai en Seoul duurt het meer dan 60 jaar voordat de geprojecteerde verdubbeling plaatsvindt. Naast onderlinge differentiatie in groeisnelheden, veranderen de steden ook in de samenstelling van het bebouwd gebied. Dit is vooral zichtbaar wanneer sprake is van een hoge mate van verdichting, die plaatsvindt in de projecties voor het merendeel van de steden.  Slechts enkele steden (bijv. Shanghai, Calcutta en Mumbai) vertonen significante spreiding van de stedelijke groei in een gefragmenteerd en uitgestrekt semi-urbaan stedelijk weefsel. Hoewel er vele overeenkomsten zijn tussen de verschillende steden in de analyse, is het nagenoeg onmogelijk gebleken om algemene trends te ontwaren, die representatief zijn voor het merendeel van de steden. Deze uitkomst lijkt de aanname te onderschrijven, dat trends in stedelijke groei niet uitwisselbaar zijn en dat iedere stad unieke groeikarakteristieken bezit, die enkel op individuele basis kunnen worden geëvalueerd.

## Impact op toekomstige overstromingsrisico's

Toekomstige overstromingsrisico's als functie van stedelijke groei, hebben zich in deze studie beperkt tot inundaties vanuit de rivier en door lokale neerslag. De mondiale impact van stormvloed op stedelijke agglomeraties aan de kust is reeds voldoende in kaart gebracht in andere studies. De schatting van rivier-gerelateerde overstromingsrisico's is gebaseerd op het GLOFRIS model, een hydrologisch model dat alle grote rivieren van de wereld omvat. Hieruit zijn kaarten met inundatiedieptes geproduceerd voor overstromingen met herhalingstijden variërend tussen de 10 en 1000 jaar. Aangezien het detailniveau tussen de inundatiedata en de landgebruikdata van de groeiprojecties behoorlijk verschilt (gridcellen van ±800m vs 30m), is er een gevoeligheidsanalyse uitgevoerd om te bepalen of deze discrepantie leidt tot significante foutmarges. Dit is slechts het geval voor een beperkt aantal steden: Shanghai, waar inundatiedieptes zeer gering zijn, alsmede Istanbul en Teheran, waarbij de blootstelling aan rivieroverstromingen marginaal is. De landgebruikdata van de groeiprojecties is relatief schematisch, waardoor het niet geschikt is voor een gedetailleerde risicoanalyse. Het bepalen van overstromingsrisico's is in deze studie daarom beperkt tot een analyse van de blootstelling, gebaseerd op de overlap van inundatiekaarten en het geprojecteerde stedelijk gebied.

De uitkomsten tonen een substantiële toename van de stedelijke blootstelling aan overstromingen bij een groei naar het richtjaar 2060. Dit is vooral het geval voor Dhaka, Ho Chi Minh City en Lahore, waar de geprojecteerde blootstelling meer dan verdrievoudigde. Wanneer puur naar de grootte van het aan overstroming blootgestelde stedelijk wordt gekeken, dan prijkt de Guangzhou-Shenzhen regio boven aan de lijst gevolgd door Calcutta, Beijing, Shanghai en Delhi, waar het betreffende gebied honderden vierkante kilometers beslaat. Daarbij geldt overigens dat de inundaties in Beijing en Shanghai relatief gering zijn, waardoor de grootte van het blootgestelde gebied wellicht overschat is. Validatie en uitbreiding van het onderzoek heeft zich gericht op met name de case study van Dhaka, waarbij de blootstelling, maar ook de geprojecteerde schades zijn bepaald op basis van een overstromingsmodel met een hoger detailniveau. Het stedelijk gebied voor de groeiprojecties is onderverdeeld in 10 dichtheidsklassen. Deze zijn gekoppeld aan een serie schadecurves. De geschatte overstromingsschades voor 2050 als functie van de geprojecteerde stedelijke groei zijn tot 6.8 maal zo groot dan die geschat voor de referentiecondities van 2004. De toename is in lijn met de verwachte groei van de blootstelling die geschat werd op 6.4 maal de blootstelling voor 2004. Een zorgelijke trend in de uitkomsten is waargenomen in steden, waar een disproportionele groei van de blootstelling aan overstromingen kan worden waargenomen in  verhouding tot de geprojecteerde groei. Groei in deze steden concentreerdt zich vooral in laaggelegen gebieden i.p.v. in gebieden die niet onderlopen. Dit is vooral het geval in Dhaka, Ho Chi Minh City en Lahore en in mindere mate in Jakarta. Een disproportionele groei van het aan blootgestelde stedelijke gebied markeert een transitie in de overstromingskarakteristieken voor die stad: het huidige overstromingsrisico wordt niet alleen groter, maar beslaat ook een relatief groter gebied op basis van de geprojecteerde groei. Integratie van overstromingsrisico's in bestemmingsplannen en groeibeteugeling van stedelijke agglomeraties lijkt daarom een belangrijke maatregel aangezien ongecontroleerde groei leidt tot een substantiële toename van de risico's. Aan de andere kant blijken er ook steden te zijn, waarbij versnelde groei met name plaats vindt in veilige gebieden. Dit is het geval in Shanghai, Seoul en Mexico City. Blijkbaar hebben deze steden een "natuurlijke drang" tot overstromingsbestendige groei. Het terugdringen van overstromingsrisico's zou zich in dit soort steden wellicht meer moeten richten op traditionele beschermingsmaatregelen. Als laatste is er ook onderzoek gedaan naar de gevoeligheid van steden t.o.v. de frequenties van rivieroverstromingen en de daarbij behorende inundaties. Substantiële vergroting van het overstroomde stedelijke gebied bij lagere frequenties kan als een proxy indicator worden gezien van de gevoeligheid van steden voor hogere rivierafvoeren als gevolg van klimaatsverandering. Steden die hiervoor bovenmatig gevoelig zijn betreffen Lagos, Mexico City en Seoul.

De methodiek voor de evaluatie van overstroming als gevolg van lokale regenval is beperkt gebleken tot het gebruik van een reeks proxy indicatoren. De vereisten voor het gebruik van geavanceerde 1d2d hydrodynamische modellen blijken dusdanig hoog, dat toepassing niet haalbaar is gebleken voor de uitgestrekte stedelijke gebieden, die als case studies hebben gediend. Als alternatief is daarom gekozen voor een reeks indicatoren die inzicht geven in de drainage condities op macro-, meso- en microschaal. Een aanname daarbij is dat de drainagekarakteristieken sterk gekoppeld zijn aan het landgebruik. Op macroschaal zorgt stedelijke groei voor wijziging van de infiltratiecapaciteit binnen de stedelijke voetafdruk. Om karakteristieke veranderingen in het patroon van het bebouwd oppervlak te meten, zijn op mesoschaal de fractale dimensie en de open landverdeling gebruikt als indicatoren. Deze drukken de mate van fragmentatie en perforatie van het stedelijk weefsel uit, wat een maatstaf is voor de beschikbare ruimte voor waterberging. Dit is vooral van belang bij extreme regenval. Op microschaal is een kwantitatieve analyse gemaakt van de landgebruiktransities, waarbij bijvoorbeeld is gekeken of de groei van hoog stedelijk gebied gebeurt door de conversie van landbouwgebied of door verdichting van voorsteden. Na evaluatie van deze indicatoren, zijn deze op semi-kwantitatieve wijze gecombineerd om tot een prognose te komen voor de afzonderlijke stedelijke agglomeraties.

Steden waarvoor de uitkomsten een relatief positief beeld schetsten zijn Ho Chi Minh City, Jakarta en Mumbai. Voor deze steden leiden groeiprojecties niet overwegend tot een uitgestrekt en aaneengesloten hoogstedelijk gebied. Met name in Ho Chi Minh City tonen de groeiprojecties een hoge mate van fragmentatie van het stedelijk weefsel, waardoor voldoende capaciteit voor waterberging gerealiseerd kan worden. Ook in Jakarta en Mumbai blijkt dit het geval, hoewel voor Mumbai geldt dat fragmentatie vooral optreedt in de perifere delen, die ver weg liggen van het schiereiland waarin het stadshart is gevestigd. Steden met relatief lage scores in deze evaluatie zijn Karachi en Istanbul, die vanuit een relatief hoge dichtheid van het huidig stedelijk weefsel verder verdichten en zeer compact georganiseerde groei vertonen. Dit laatste zorgt voor zeer lage gemiddelde infiltratiecapaciteit. Agglomeraties waar een disproportioneel sterke daling is geconstateerd van 25% of meer, zijn Guangzhou-Shenzhen, Manila, Mexico City en Dhaka. Vanzelfsprekend worden in deze evaluatie de verschillende neerslagkarakteristieken tussen de steden buiten beschouwing gelaten, terwijl die substantieel kunnen verschillen tussen steden als bijv. Manila en Teheran. De uitkomsten richten zich echter veel meer op de verwachtte relatieve verschillen; m.a.w. verandert de relatieve drainagecapaciteit fundamenteel onder invloed van de geprojecteerde stedelijke groei. Voor Karachi en Istanbul lijkt dit inderdaad sterk het geval te zin. Ook voor Dhaka en in mindere mate Delhi, Guangzhou- Shenzhen en Teheran wordt de bergings- en infiltratiecapaciteit van regenwater disproportioneel lager.

De uitkomsten voor rivieroverstromingen en overstromingen door hevige regenval zijn met opzet niet samengevoegd in een gecombineerd resultaat. Naast een verschil in ordegrootte van de problematiek is ook de benadering en toegepaste metrik onderling zo verschillend dat het combineren van uitkomsten niet relevant is. Desondanks lijkt stedelijke groei voor Dhaka, Guangzhou-Shenzhen en Lahore te zorgen voor een disproportionele toename in de kwetsbaarheid voor overstromingen in beide domeinen. Vooral in deze steden kan planning een grote rol spelen bij het beperken van toekomstig overstromingsrisico's.

## Toepassingen in andere domeinen

De toepassing van stedelijke groeiscenario studies is uiteraard veel breder dan enkel het bepalen van toekomstige overstromingsrisico's. Om dit te illustreren zijn twee alternatieve studies gedaan, die voornamelijk dienen als basisimplementatie ofwel proof-of-concept. De eerste toepassing heeft zich gericht op de toename van verontreiniging in watergangen als gevolg van de onstuimige groei van sloppenwijken en lozing van afval in het oppervlaktewater. Ook hiervoor is ruimtelijk inzicht in de groeipatronen van belang voor het bepalen van de locatie en de potentiele toename van de verontreiniging. De groeiprojectie voor Lagos is hierbij gebruikt als casus. De uitkomsten tonen aan dat voor een aantal watergangen en afvoerpunten een disproportionele toename van verontreiniging optreedt. Dit vanwege de groei van sloppenwijken in de directe nabijheid van een beperkt aantal watergangen. Deze toename kan ingrijpende gevolgen hebben voor de waterkwaliteit en ecosysteem van specifiek regio's in de delta, waarin Lagos is gevestigd. Ook de economisch consequenties zijn mogelijk bovenmatig groot aangezien een aanzienlijke gemeenschap economisch afhankelijk is van de delta. Voor het tweede domein, is een studie gedaan naar de gevolgen van stedelijke groei voor het hitte-eiland en de resulterende veranderingen in neerslagpatronen voor Mumbai. Door gebruik te maken van een 3d-atmosferish model op mesoschaal, waarbij veranderend landgebruik is gebruikt als één van de parameters, blijkt dat historische buienreeksen toenemen in intensiteit, incl. die van de piekbuien. Aangezien de uitkomsten afhankelijk zijn van de specifieke geografische condities en de groeipatronen van Mumbai, zijn de uitkomsten niet representatief voor andere steden. Desondanks toont de studie aan dat stedelijke groei ook een effect kan hebben op de neerslagcomponent binnen de risicoketen en niet slechts op de blootstelling of de gevolgen; stedelijke groei kan leiden tot verhevigde neerslag. Deze twee toepassingen tonen aan dat ruimtelijk expliciete stedelijke groeiscenario's als basis kunnen dienen voor een groot scala aan milieueffect- en omgevingsstudies. Deze kunnen als referentie dienen voor het meten van de gevolgen van maatregelen en bieden een alternatief voor een statische benadering, waarbij de huidige stedelijke condities als referentie worden genomen voor lange termijnstudies.

## Conclusies

Om de integratie van alternatief stedelijk groeibeleid, strategieën en zoneringsplannen beter te kunnen faciliteren, dient het in de studie gebruikte model verder te worden ontwikkeld. Tevens moet de gebruikersvriendelijkheid en flexibiliteit verder worden vergoot om toepassing buiten het huidige onderzoekdomein te vereenvoudigen. Op dit onderzoeksgebied, is het vergelijken van de effecten van stedelijke groei met die van toekomstige klimaatverandering van essentieel belang. Zo kan eindelijk inzicht worden verkregen in hoe de gevolgen van deze ontwikkelingen zich tot elkaar verhouden. Voor rivieroverstromingen zou dit relatief eenvoudig moeten zijn zodra die overstromingsdata beschikbaar is.

Om de impact van deze studie te vergroten dient er een adequaat platform te worden gevonden buiten de wetenschappelijke wereld. Hoewel uitgebreide stedelijke netwerken voorhanden zijn die zich richten op klimaatadaptatie, natuurrampen en van duurzame ontwikkeling, hebben deze vaak een focus op ondersteuning van individuele steden. Ruimte voor grootschalig vergelijkend onderzoek is meestal geen prioriteit. Desondanks zijn dit soort studies essentieel voor een beter begrip van de positie van het stedenbouwkundige en regionale ontwerp in het beperken van toekomstige klimaat-gerelateerde rampen en de prioritering daarvan in de stedelijke ontwikkeling van die steden waar ongecontroleerde groei tot een disproportioneel hoge kwetsbaarheid leidt.

# Contents

Summary                                                                    v

Samenvatting                                                               xi

Contents                                                                  xix

Figures and Tables                                                       xxiii


1.  Background                                                             1

1.1  Introduction                                                          2

   1.1.1  Typical challenges                                               5

1.2  Urban growth: towards the building blocks                            5

   1.2.1  Modelling urban growth                                          9

1.3  Urban growth and scenario development                               16

1.4  Urban growth and flood risk                                         18

   1.4.1  Cities and riverine flood impacts                              18

   1.4.2  From imperviousness to surface runoff                          20

   1.4.3  Towards the drained city                                       22

1.5  Flood vulnerability assessment                                      24

1.6  Converging to the scope                                             25

1.7  Reader's guide                                                      25

1.8  Embedded research projects                                          27

2.  Research Questions and Methodology                                   29

2.1  From state-of-the-art to research questions                         30

2.2  Hypotheses                                                          31

2.3  Research Methodologies and skills                                   34

2.4  Methodological considerations                                       35

   2.4.1  Urban growth model and scenarios                               35

   2.4.2  Riverine flooding                                              36

   2.4.3  Pluvial flooding                                               37

   2.4.4  Pre- and Post-processing                                       37

2.5  Peripheral topics                                                   39

2.6  Originality, innovation and potential impact                        40

2.7  Selection of case studies                                           40

3.  Memetic algorithm optimised urban growth model                       43

3.1  Introduction                                                        44

3.2  The Case Study                                                      46

   3.2.1  Beijing                                                        46

3.3  Data and model setup                                                      48
  3.3.1  Model refinement: Sequential approach                        53
  3.3.2  Automated Calibration                                        55
3.4  Outcomes                                                                   59
  3.4.1  No Free-Lunch                                               63
3.5  Projections                                                                64
3.6  Discussion                                                                 65
3.7  Conclusion                                                                 67
4.  Urban growth projections                                                    69
4.1  Introduction                                                               70
4.2  BAU for urban growth                                                       70
4.3  Historic and projected urban growth                                        72
  4.3.1  Urban composition                                          77
4.4  Growth potential and characteristics                                       79
  4.4.1  Spatial constraints                                         79
  4.4.2  Growth attractors                                           81
4.5  Conclusions                                                                84
5.  Future riverine flooding in megacities                                      87
5.1  Introduction                                                               88
5.2  Urban growth and floods                                                    89
  5.2.1  Datasets                                                     91
  5.2.2  Assessing the urban flood extent and depth distribution     94
5.3  Validation                                                                 95
  5.3.1  Initial setup                                               95
  5.3.2  Upscaling                                                    96
  5.3.3  Downscaling                                                  96
  5.3.4  Additional validation methods                               97
  5.3.5  Validation results                                          98
    5.3.5.1  Upscaling                                      98
    5.3.5.2  Downscaling                                    99
    5.3.5.3  Additional Data Sources                       100
  5.3.6  Conclusions from the validation                            101
5.4  Outcomes                                                                  103
  5.4.1  Urban flood extent                                         104
    5.4.1.1  Flood extent differentiation                  111
  5.4.2  Flood depth distribution                                   111
5.5  Evaluation and Conclusions                                               114

5.5.1  Ranking in relation to coastal flooding                                116

5.5.2  Consequences for urban flood risk management              117

5.5.3  Responding to increasing urban flood risk                        118

5.6  Extending the outcomes: CC-sensitivity                                 120

5.7  Discussion                                                                                124

6.  Assessing the effects of urban growth on urban drainage       127

6.1  Introduction                                                                             128

6.2  Operationalising future drainage performance through ISR   131

6.2.1  Case Beijing: extensive soil sealing due to concentric urban develop-
ment                                                                                               132

6.2.2  Comparing the drainage performance                              138

6.2.2.1  Macro level assessment                                                 138

6.2.2.2  Meso level assessment                                                  139

6.2.2.3  Micro level assessment                                                 141

6.3  Outcomes: Towards a Sponge City                                       144

6.4  Discussion                                                                             146

6.4.1  Policy options                                                                    147

6.5  Conclusions                                                                           148

7.  Adding depth: Estimating flood damages in Dhaka                  151

7.1  Introduction                                                                            152

7.2  Dhaka case study                                                                   153

7.2.1  General characteristics of the city                                   153

7.2.2  Urban growth                                                                     154

7.2.3  The 2004 flood                                                                   156

7.3  Urban growth model, flood model and damage model         156

7.3.1  Flood model                                                                       156

7.3.2  Flood damage model                                                         157

7.4  Scenarios                                                                                158

7.5  Outcomes                                                                               160

7.5.1  Comparison of flooding                                                     161

7.5.2  Comparison of flood damages                                          161

7.6  Interpretation                                                                         162

7.7  Discussion                                                                              164

8.  Further explorations                                                                  165

8.1  Urban growth modelling and implications on water supply and sanitation
planning                                                                                         166

8.1.1  Introduction                                                                        166

| | |
|---|---|
| 8.1.2 Lagos | 167 |
| 8.1.3 Methodology and outcomes | 168 |
| 8.1.3.1 Urban Growth Model and projections | 168 |
| 8.1.4 Watershed Delineation and pollution loads | 170 |
| 8.1.5 Discussion | 172 |
| 8.2 Urban growth and microclimate | 172 |
| 8.2.1 Introduction | 172 |
| 8.2.2 WRF-ARW Model | 173 |
| 8.2.3 Mumbai case-study with future urbanisation | 174 |
| 8.2.4 Outcomes | 175 |
| 8.2.5 Discussion | 176 |
| 9. Towards an argument | 179 |
| 9.1 Answering the RQs and testing hypotheses | 180 |
| 9.2 Conclusions | 186 |
| 9.2.1 Riverine flooding | 186 |
| 9.2.2 Pluvial flooding | 188 |
| 9.2.3 Consequences for flood risk management | 190 |
| 9.3 Recommendations | 190 |
| 9.3.1 Urban growth model | 190 |
| 9.3.2 Riverine flood risk assessment | 191 |
| 9.3.3 Pluvial flood risk assessment | 192 |
| 9.3.4 Additional flood hazards | 193 |
| 9.3.5 Scenarios | 193 |
| 9.3.6 Assessment | 194 |
| 9.4 Discussion | 195 |
| 10. Bibliography | 199 |
| | |
| Appendix A: Urban growth and riverine flooding | 221 |
| Appendix B: Pluvial flooding | 303 |
| Acknowledgements | |
| Curriculum Vitae | |
| Publications | |

# Figures and tables

## List of Figures

**Figure 1**: Different components of this study and their interactions 34

**Figure 2**: Overview of all cities at same scale (top row): Lagos, New Delhi, Tehran, Cairo and (bottom row) Mumbai, Dhaka, Beijing, Guangzhou-Shenzhen, Kolkata, Ho Chi Minh City. 38

**Figure 3**: Typical setup for a LULC change model, including the feedback mechanism for the calibration 44

**Figure 4**: Estimated urban development over 1995-2005 based on Landsat TM/ETM data 46

**Figure 5**: Weight distribution for the LULC transition between grassland and low density built-up areas as a function of the distance to the main road network 51

**Figure 6**: Exploration (left) and exploitation (right) in a GA 55

**Figure 7**: Implementation of the MA into the calibration and validation sequence. 58

**Figure 8**: Mean and $5^{th}$-$95^{th}$ percentiles for the observed maximum MMS values using GA-enhanced (left) and MA-enhanced (right) calibrations. 60

**Figure 9**: Observed distribution of the MMS using GA-optimized (left) and MA-optimized (right) calibrations. 61

**Figure 10**: Problem of direction (left) and step-size (right) in local searches 63

**Figure 11**: Projected LULC distributions for Beijing, including details on the urban areas produced for 2060. 64

**Figure 12**: Projected LULC distribution for Beijing using the 2-stage model, including details on the urban areas produced for 2060 (bottom).2060. 65

**Figure 13**: Estimated 1990 and 2010 urban footprint for the Guangzhou-Shenzhen region as well as the projected urban growth between 2010-2060. 73

**Figure 14**: Estimated urban footprint and growth 74

**Figure 15**: Urban landscape composition for Guangzhou-Shenzhen in 2010 (left) and 2060 (right) 78

**Figure 16**: Derived weights distribution of slope and elevation for Seoul 82

**Figure 17**: Derived weights distribution of distance to infrastructure for Ho Chi Minh City 83

**Figure 18**: Built-up areas in the original 30m cell grid sized dataset (left) and resampled to 30 Arc Seconds (right). 96

**Figure 19**: Flood extent superimposed on the GW elevation data (left) and the SRTM adjusted extent superimposed on the SRTM data (right) 97

**Figure 20**: Mean, maximum and minimum flood extent as a function of urban growth 100

**Figure 21**: Dhaka urbanised flood extent as a function of urban growth 102

**Figure 22**: Development of the estimated urban flood extent over time as a function of urban growth   105

**Figure 23**: Growth rates of flood prone against flood secure urbanised areas for Lahore and Shanghai   109

**Figure 24**: Flood extent over time for different return periods for Lagos (left) and Jakarta (right)   110

**Figure 25**: Estimated flood depth distribution for Guangzhou-Shenzhen for different years and return period.   112

**Figure 26**: Comparison between estimated urban growth rates within and outside the average flood extent for the interval 2015-2060.   114

**Figure 27**: Growth of the urban flood extent for 2010 over increasing return periods   121

**Figure 28**: Urban development in a section of Dhaka, showing the area in 2001 (top), 2008 (centre) and 2015 (bottom). Photo's courtesy of Google Earth™   129

**Figure 29**: Estimated ISR distribution for Beijing in 2005 (left) and 2060 (right)   133

**Figure 30**: Fractions of different LULC-classes and associated ISRs for Beijing   133

**Figure 31**: Modelled and extrapolated projections of built-up (left) and open areas (right) in Beijing   137

**Figure 32**: Estimated FD for 2015 (left) and the 2060 (right)   140

**Figure 33**: Progression of ratio of open land with declining (top-left) and increasing (top-right) trends as well as stable (bottom-left) and irregular (bottom-right) trends.   142

**Figure 34**: Generalized proportionality built-up areas 2015-2060   143

**Figure 35**: Eastern and Western Dhaka based on drainage separation   154

Figure 36: Dhaka urban development between 1990 and 2005 (left) and eastern Dhaka drainage system (right).   155

**Figure 37**: Aggregate damage curves for different densities   157

**Figure 38**: Growth characterisation (left) and resulting distribution of built-up areas including flood extent (right).   158

**Figure 39**: Terrain map (left), estimated flood depth (centre) and flood damages (right) for the 2050 scenario.   160

**Figure 40**: Comparison of overland flow for different scenarios   161

**Figure 41**: Division of damages over urban footprint for 2004 (left) and 2050 (right).   163

**Figure 42**: Children in Makoko slum in Lagos. Source: NOVA Next   167

**Figure 43**: Google Earth ™ aerial photo of the Iwaya neighbourhood (left) transitioning into the lagoon oriented Makoko slum (right).   168

**Figure 44**: Observed land cover map of 2010 and simulated land cover maps for 2035 and 2060   169

**Figure 45**: Urban growth statistics per urban LULC class for 2000, 2010 (observed), 2035 and 2060 (projected)                                                                  170

**Figure 46**: Sanitary Pollution Loads in slums of Lagos - 2010, 2035, 2060                                   171

**Figure 47**: Urban growth of Mumbai metropolitan area 1990, 2005 (observed) and 2035, 2060 (projected)                                                                                             174

**Figure 48**: Quintile-quintile plots of rainfall intensities (left) and estimated present and future rainfall frequencies (right)                                                        175

**Figure 49**: Total rainfall accumulations (mm) during the 2007 July rainfall event simulation using the 2005 (left) and projected 2060 (right) LULC map. The prevailing surface wind direction is marked by the arrow.                                                      176

## List of Tables

**Table 1**: Steps and parameters used for calibration                                                    58

**Table 2**: Summary of the main characteristics of the applied GA and MA schemes   59

**Table 3**: Mean, 5th, 95th percentiles and resulting range of MMS values after 96 and 32 iterations for the GA and MA optimized calibrations, respectively.            61

**Table 4**: Confusion matrix and kappa index for the 1995-2010 transitions. The projected LULC cell changes are in the rows, and the observed cell changes are in the columns.                                                                                                      62

**Table 5**: Doubling periods                                                                                         76

**Table 6**: Dominant geographic growth constraints (indicated in black) and growth statistics                                                                                                  80

**Table 7**: Estimated flood extent for Seoul for the initial setup and alternative procedures                                                                                                     103

**Table 8**: Estimated flood extent for Shanghai for the initial setup and alternative procedures                                                                                                  103

**Table 9**: Top and bottom ranking based on urban flood extent for 2015 and 2060                                                                                                              104

**Table 10**: Comparison of the ranking of flood exposed cities                                      117

**Table 11**: Ranking based on the estimated growth ratio                                            123

**Table 12**: Ranking of cities based on ISR for 2015 and 2060                                    138

**Table 13**: Qualitative assessment of the cumulative and scale dependent indicators                                                                                                            145

**Table 14**: Model combinations for scenarios                                                          160

**Table 15**: Total damage in study area for different scenarios                                    162

**Table 16**: Estimated pollution loads                                                                      171

**Table 17**: Ranking of cities exposed to coastal, riverine and pluvial flooding        187

# Glossary

**BAU:** business as usual

**CA:** cellular automata

**CC:** climate change

**COAMPS:** Coupled Ocean/Atmosphere Mesoscale Prediction System

**DDC:** depth-damage curve

**DEM**: Digital Elevation Model

**e.g.:** exempli gratia, meaning "for example"

**et al.:** et alii, meaning "and others"

**FD:** fractal dimension

**GA:** genetic algorithm

**GAS:** geographic automata system

**GDEM**: Global Digital Elevation Map

**GIS:** geographic information system

**GLOFRIS:** Global Flood Risk with IMAGE Scenarios

**GW:** Global Watershed

**ibid:** ibidem, meaning "the same place"

**ICLEI**: International Council for Local Environmental Initiatives

**i.e.:** id est, meaning "that is"

**ISR**: impervious surface ratio

**IDF**: intensity-duration-frequency

**IWM:** Institute of Water Modelling

**LIDAR:** Light Detection and Ranging

**LULC:** land use and land cover

**MA:** memetic algorithm

**MMS:** minimum mean similarity

**RS:** remote sensing

**SRTM**: Shuttle Radar Topography Mission

**RS:** remote sensing

**UHI:** urban heat island

**USGS:** United States Geological Survey

**WRF:** Weather Research and Forecasting Model

# 1. Background

## 1.1 Introduction

In recent years, various studies provide evidence for an increased future vulnerability of many of the world's cities to flood impacts (e.g. Aerts et al, 2014; Huq et al, 2007; Jha et al, 2012; Merz et al, 2010). This often seems a result of climate change-induced trend changes and increased variability in precipitation, which changes the distribution of flood events. More variability as well as an increased likelihood of extreme rainfall is expected particularly in regions that already suffer from a periodical abundance of precipitation (e.g. Milly et al, 2002). This in turn is likely to cause more frequent and severe pluvial and fluvial floods and subsequent impacts (e.g. Stern, 2007).

Yet, apart from the consequences of climate change, the perceived increased flood risk is also a consequence of other drivers; one of the prime factors being the increased susceptibility to flood impacts in many of the world's urbanized areas caused by a process of unprecedented urban expansion over the last century (UN, 2014; Fuchs, 1994). This results in a massive allocation of people and assets in flood prone areas thus increasing the potential impact from future flooding (both in frequency and intensity). Consequently, the framework for flood risk management requires reconsideration. Protection schemes based on static design floods are facing a new perspective in which "stationarity is dead" (Milly et al, 2008) and in which continuous change and adaptation to future climate related hazards is proposed.

Arguably though, the insights into the consequences of flooding on the urban environment lack both a formal definition and method for sound impact assessment (e.g. Wind et al, 1999; Thieken et al, 2005). This weakens the decision framework for proposed measures. While the notion that the complexity of the climate system might lead to inherently uncertain forecasts of future trend changes becomes accepted both in and beyond the scientific community (e.g. Haasnoot et al, 2013; Füssel, 2007), the issue of extensive urban growth in relation to future climate related impacts, remains underexposed. The development and integration of climate change scenarios has become common practise in future assessments (Hall et al, 2005; Nakićenović, 2000). The explicit formulation of future urban growth projections is often neglected or rendered as an issue that in terms of complexity is regarded intractable (e.g. Schreider, 2000; ). That means that while a probabilistic approach to future climate change scenarios is in some cases already disqualified by the scientific community (e.g. Scoones, 2004), proper attempts to develop even business-as-usual (BAU) scenarios for urban development have in many cases not been developed. This creates a mismatch between the advances in thinking about climate change and the dynamics of one of the most vulnerable receptors of climate change: urbanised areas (e.g. (UN-HABITAT et al,

2011). Where for the development of climate adaptation strategies scenarios are used that easily cover 50 years or more, too often future assessments are based on current urban conditions. So, one could state that 'stationarity is not dead', on the contrary: 'immutability is still common'. Although these uncertainties have been identified (e.g. Merz et al, 2010) little progress has been made to bridge this gap. If scenarios have been used to incorporate future changes in water management, they rely on statistical extrapolations or parameter changes (.e.g Semadeni-Davies et al, 2008) instead of spatially explicit land use and land cover (LULC) changes.

Nevertheless, the impacts of urban growth as a driver for increasing future climate related impacts are widely accepted, especially in relation to flood risk (e.g. Jha et al, 2012). This holds for coastal (e.g. Nicholls et al, 2008), fluvial (e.g. Mitchell, 2003) as well as for pluvial floods (e.g. Weng, 2001) and for all components that constitute risk (e.g. Cardona et al, 2012; Zevenbergen et al, 2011): hazard (e.g. Carlson et al, 2000), exposure and vulnerability (or sensitivity). In coastal flooding, future assessments have been primarily based on statistical extrapolations of urban key indicators (e.g. population, asset value) in relation to increased exposure due to climate change-induced sea level rise (Hallegatte et al, 2013). Advancements in urban growth affected fluvial flood risk have been limited to either single case studies (Moel et al, 2010) or relatively schematic approximations (e.g. Jongman et al, 2012). This limited scope extends to pluvial flooding, where emphasis has been on retrospective estimation of the drainage characteristics (Yang et al, 2005) often focussing on basin scale (e.g. Shi et al, 2007; Bruin 2000).

Yet, we are living in an age of cities, where rapid urban development is currently changing not only the socio-economic but also the biophysical characteristics of many regions located in the world's deltas or further upstream adjacent to major rivers at a massive scale (e.g. Angel et al, 2005). Apart from the impacts on the water cycle (e.g. Huong et al, 2013) this process also changes the risk profile of vast regions, where cities can be considered the economic and demographic hubs. Integration of urban development projections should therefore be a standard ingredient in future flood risk assessment as well as the development and evaluation of flood adaptation strategies and measures. This is especially prudent since from all natural hazards, floods cause the majority of damages (Munich RE, 2005). A better outlook on how rapidly evolving metropolitan areas perform in the future is therefore essential for moving towards a more proactive management of flood risk instead of fixing 'past mistakes' that could be the product of ignorance and subsequent inaction.

Integrating future urban development in flood risk touches upon an important issue: local specificity. Floods are local phenomena; apart from their frequency and amplitude, impacts of inundations are largely defined by the area they cover: the flood extent. In the case of cities, the spatial attributes of a flood interact directly with those of its urban receptor. This notion adds a requirement for the application of urban development scenarios in future flood risk assessment and management: spatial explicitness. Only if urban growth scenarios are geographically bounded, they can express both the urban differentiation within and across cities as well as their interaction with coastal, fluvial and pluvial floods. This especially holds for cities in the developing world that witness a substantial growth differentiation that sometimes defies common assumptions (e.g .Cohen, 2004).

Ironically, spatial models that attempt to explain urban growth exist already since the early 1960s and matured into sophisticated LULC change models together with the widespread availability of remote sensing data and computational capacity in the early 1990s (e.g. Batty 2007; Benenson et al, 2004). Up till now, applications beyond the domain of geography and computational sciences have been limited. This seems remarkable, since their potential in the domain of climate adaptation, including flood risk management seems extensive. Cities have been identified as key-actors in the development and implementation of climate change adaptation strategies (ICLEI, 2003) and are as such united in extensive networks in which they exchange knowledge, share resources and develop common agendas, goals and strategies (e.g. Rockefeller Foundation, 2013; UNISDR, 2010; ICLEI, 2003). Despite these efforts and initiatives, the question remains how cities can develop effective policies without a baseline scenario to which the effectiveness of future plans can be evaluated. The development of spatially explicit urban growth scenarios, based on extrapolations of past trends (i.e. business-as-usual scenarios) should be a top priority to assess the potential impact of future interventions that aim to mitigate or adapt to future natural hazards, including floods.

Before such questions can be answered, a relevant question is how spatially explicit growth scenarios affect the future outlook of cities in terms of natural hazards, and particularly of floods. Continuation of past growth trends might for some cities lead to rapidly increasing flood exposure, while for others the projected changes have few implications. Some cities might perform particularly badly in relation to riverine flooding, while for others the main challenge might be urban drainage. Possibly clusters of cities can be identified, based on similarities in how their pathways towards future flood risk development. Such answers could lead to alternative prioritizations, different approaches or at least contribute in the discussion of a more flood resilient urban development. Furthermore, better insight might be gained in how the contribution of urban growth to future flood risk compares to the projected impacts of climate change.

## 1.1.1 Typical challenges

The issue thus becomes to develop a set of spatially explicit urban growth scenarios, assess future flood risk and to develop a set of metrics that effectively express the future performance and to evaluate how that performance compares to the present conditions. While these questions seem straightforward, they are founded on a set of implicit assumptions that need to be investigated before even an attempt can be made to develop a suitable approach for these challenges. If for instance the urban development scenarios are based on extrapolations of past spatial development trends, a more formal definition what spatial trends actually need to be formulated. To assess future flood risk, the urban growth scenarios need to be expressive enough to interact with flood models or sets of inundation maps that in turn represent different types of floods (e.g. coastal, fluvial, pluvial). The desire to compare outcomes between and across cities, sets requirements for a uniform approach that allows only limited differentiation in for instance the datasets. Finally, to overcome the limitations of some of the past studies and to develop relevant and robust conclusions, the assessment should cover a relatively large number of case studies and span a sufficiently long period.

To further elaborate on some of the ramifications of these assumptions and to further specify these broad goals, the issues are described in further depth. First a more in-depth description of the main focus of this research is provided: urban growth. Then, a section of developing the specifications for the urban growth scenarios and the LULC change model required to produce these, are given. The scenarios can be regarded as a spatiotemporal foundation for a range of environmental impact assessment.

## 1.2 Urban growth: towards the building blocks

The year 2008 marked a turning point in global demographics: more than 50% of the world's population lived in urban areas (UNFPA, 2007). This turning point has been preceded by decades of unprecedented urban growth that is only expected to continue in the coming decades. In developing countries, by 2030 the urban population is expected to rise to almost 4 billion inhabitants (UN, 2004), a 100% growth within about 30 years.

This growth also marked the rise of megacities with populations exceeding 10 million inhabitants, which first started in the early 1950s with the urban agglomerations of New York and Tokyo (UN, 2015). Currently the world is hosting 29 megacities, of which the Tokyo-Yokohama metropolitan area is considered the largest with a population close to 40 million inhabitants (ibid). Typically, every decade around 6 new megacities emerge. In 1980, there were only 5 megacities. This number steadily grew to 10 in

1990, 17 in 2000 and 23 in 2010 (ibid). In 2030 the number of megacities is estimated to become 41 (ibid) with 23 Asian megacities, although that might be an underestimation given the rapid growth of many upcoming agglomerations.

Some of these figures are disputed since many factors significantly impact the census of large metropolitan areas. For instance, a significant portion of unregistered dwellers resides in informal settlements making proper registration difficult. Furthermore, often estimations are based on outdated figures. An additional factor that impacts population figures is the lack of consensus (or explicit clarity) about the analysis extent (i.e. the area-of-interest). Some numbers are based on the administrative borders (e.g. the municipal boundaries) that only cover the main urban core, while others include for instance suburbs, villages and other small pockets of urbanisation in the immediate vicinity of the urban core. For instance, the fact that top-ranked Tokyo-Yokohama is a union of two initially separate metropolitan areas already indicates that the boundaries that are used for such estimations are not always consistent. This can lead to significant differences between estimations that can sometimes differ an order of magnitude (Potere et al, 2007). As a consequence, the OECD estimates these numbers to be significantly larger since the UN figures are based on administrative units instead of addressing the actual functional regions these urban agglomerations occupy. According to the OECD, China currently already hosts 15 megacities instead of the 6 the UN identifies (OECD, 2015).

Apart from population growth, the development of megacities is mainly driven by urban economic development and the associated rural to urban migration which sadly also drivers the number of urban poor that live in many of the vast slum areas. This, in return boosts unplanned growth of cities. For cities in the developing world, UN Habitat (2007) estimates that only 5% of total urban growth is planned. The proliferation of slums is becoming the main problem associated with the expansion of cities in many developing countries. Apart from slum development, unplanned development also occurs at the higher end of residential development where land grabbing and rapid turnover from agriculture to vast new neighbourhoods (including gated communities) often occurs outside the designated areas assigned for urban expansion.

While cities are growing, the average densities of these urban areas are declining. Angel et al (2005) estimates an annual decline of urban densities of 1.7% for developing countries, resulting in a built-up area of 600,000 square kilometres by 2030. To put this in perspective: urban agglomerations in 2030 will have tripled occupation space with about a 160 square meter transformation of non-urban to urban per new resident. Within the industrialized world, these figures are less dramatic. Here, the urban population is expected to rise 11% within the next thirty years to about 1 billion inhabitants.

Occupied land is expected to increase 2.5 times with an annual decline in density of 2.2%. Individual occupation is substantially higher though, every new resident is expected to convert on average 500 square meters of non-urban into urban land (UN, 2015). This confirms the much higher amount of used square meters per capita. Overall, the global urbanized built-up area is expected to rise from 400.000 square kilometres in 2000 (about 0.3% of total land area of countries) to more than 1 million square meters in 2030 (about 1.1% of the total land area of countries).

These figures are based on statistical analysis, extrapolation of current trends as well as socio-economic pathways that have been developed for many regions in the world (e.g. O'Neill, 2015). Yet these figures do not provide insight into the physical manifestations of urban growth for different cities. Important issues like the geographical distribution of urban clusters, densification of urban centres or expansion along major infrastructure lines are not covered. Depending on the actual local conditions, urban growth manifests itself differently. To better assess such attributes, a classification is required that characterizes some of these typical patterns associated to cities. An important prerequisite of such a classification is, that it doesn't depend on cultural conceptions. For instance, the notion of a city centre might be subject to considerable differences across the world. That means that a set of metrics, criteria and classifications should be applied that are relatively robust (i.e. they are not subject to personal preferences), are focussed on the geographical aspects of urbanisation patterns. Although this might add a limitation, the choice of spatial metrics to characterise urbanisation patterns is daunting (O'Neill et al, 1999; Mcgarigal, 1995). Angel et al (2007) presented a set of metrics they initially developed to assess urban growth in a comprehensive retrospective study (Angel et al, 2012; Angel et al, 2005). These characterise urban areas based on density relations between built-up areas. The classification consists of five urban classes to characterise cities:

- *Main urban core*: contiguous groups of built-up pixels which at least 50% of the surrounding neighbourhood within a area of 1 km2 is built-up;
- *Secondary urban core*: pixels not belonging to the main urban core with 1km2 neighbourhoods consisting of 50% built-up area;
- *Urban fringe:* pixels with 1 km2 neighbourhoods that are 30-50% built-up;
- *Ribbon development*: semi-contiguous strands of built-up pixels that are less than 100 meters wide and have 1 km2 neighbourhoods that are less than 30% built-up;
- *Scattered development*: built-up pixels that have neighbourhoods that are less than 30% built-up and not belonging to the ribbon development

Due to the focus on densities, the metrics are particularly aimed at raster representations of urban areas instead of vectorised maps. They aim at land cover rather than on land use (Comber et al, 2005) which makes the classification particularly useful in the context of rasterized maps derived from remote sensing data.

One of the impacts on land transformation due to the growth of metropolitan areas are the extensive areas covered by urban sprawl. Conceptually, the concept of sprawl has not been clearly defined, but defining sprawl in terms of built-up area densities is an accepted procedure (UNFPA, 2007). Especially in the United States, sprawl dominated the urban growth of many cities over the last decades (e.g. Bruegmann, 2005). This low-density development creates a relatively large footprint per capita which is further amplified by an increased extent of infrastructural and utility-lifeline networks. To extend the classification the spatial distribution of urban areas with set spatiotemporal classes that characterise different types of urban growth, an additional set of metrics has been defined (Angel et al, 2007) that differentiates between:

- *Infill*: new development within remaining open spaces in already built-up areas. Infill generally leads to higher levels of density increases contiguity of the main urban core.
- *Extension*: new non-infill development extending the urban footprint in an outward direction.
- *Leapfrog development*: new development not intersecting the urban footprint leading to scattered development.

Also these classes can be easily derived from raster based maps, but require maps from different consecutive moments in time. To acquire such maps, typically land-cover data is derived from remote sensing imagery, i.e. satellite photos based on multispectral waves (e.g. infrared). Currently many sources are available from different sensors, e.g. ASTER, IKONOS, Landsat, MODIS. These all cover spectral bands, spatial scales and pixel sizes. Differentiation in land-cover is based on different levels of reflectance of the photographed objects. Different soil types are associated with different levels of reflectance and thus allow for classification (e.g. grassland, water, barren land, etc.). Through a semi-automated classification process (i.e. supervised classification), these features can be extracted into predefined land-cover classes. The applied classification scheme often differs per country. The National Land Cover Data (2001) classification applied in the USA (USGS, 2007) currently uses 35 land-cover classes from which 4 different classes are used to identify urbanized areas (Developed Open Space, Developed Low Intensity, Developed Medium Intensity and Developed High Intensity). The Corine Land Cover classification used within the EU applies 44 land-cover classes including

11 different classes for urbanized areas. This means that exchange and comparison of land-cover data is not necessarily straightforward. Yet, remapping procedures for land-use classes are generally available.

The described urban classifications, metrics and land-cover classification provide a basic framework to interpret urban growth as a process of land-cover change over time. Urban growth can now be characterized and quantified using a structured approach with uniform metrics, which is essential for comparative studies.

## 1.2.1 Modelling urban growth

The first formal models explaining urban growth, spatial distribution of settlements, land-use, population, etc. were derived from theories within spatial economy. These models focus on equilibrium states; generally stable urban structures in which the spatial distribution of resources (capital, labour and materials) is distributed in the most efficient manner. During this period many basic geographic laws were discovered that still operate within the contemporary debate on urban growth. Central place hierarchy (Weber, 1909), power distribution of settlements (Allen, 1954) and equilibrium states defining e.g. property prices (Alonso, 1964) are still at the heart of especially urban regional economics (e.g. McCann, 2001).

From the early 1960s on, these static equilibrium models were gradually replaced by dynamic models claiming that one of the main forces behind urban dynamics were positive feedback loops (e.g. Forrester, 1969). This notion originated from the studies in natural sciences (i.e. biology, ecology) where the intuition appeared that positive feedback loops dominate temporal intervals within the systems dynamics. A typical example for this is predator-prey relationships (e.g. Berryman, 1992), where oscillating population sizes follow the availability of resources (e.g. grass, rabbits and wolves). In other words, the populations exploit short term opportunities that are amplified and give rise to specific trend within the bounds of a temporal interval (e.g. rapid increase of the rabbit population due to the ample availability of grass). When translated to urban systems, this means that urban growth and relocation of resources were reinforced by e.g. economic opportunity, thus creating a momentum for different behavioural patterns over time. From a modelling perspective it is important to realize that all of these models are setup as sets of equations; they do not use actual geographic data either as input or output. Only during the introduction of geographic information systems (GIS) and remote sensing (RS) actual "geosimulation" started to appear.

The introduction of geographic information systems (GIS) introduced an important paradigm into urban modelling: the cell as a representation of an urban unit (Benenson et al, 2004). These became a mainstream feature with the introduction of raster based

GIS and remote sensing imagery. Furthermore, GIS introduced the first cell-based urban growth models adding cell states (e.g. a land-use class) to each cell. Cell states could change over time, thus introducing a formal characterization of geographic urban growth models. Many of these models applied some regression method to forecast urban growth based on time-series data on urban spread (Donnelly et al, 1964) and are still used today (Pijanowski et al, 2002).

Computer science in the meantime produced a very similar framework called Cellular Automata (CA) initially as a theoretical model for self-reproducing machines (Turing, 1936) and later formalized as a working model (Von Neumann, 1951). The formal definition of a CA shows many aspects of the concepts later to be found in contemporary urban growth models. Each automaton (i.e. cell) is defined by a set of discrete states. A set of transition rules changes the state of a cell at the next time step and depend on the cell's state at the current time step. All cells are placed on an a n-dimensional lattice which creates a topological relation between cells. A grid of automata becomes a CA when the set of inputs is defined by the states of neighbouring cells. In other words: a cell state within the next time step is defined by the neighbouring cells states. Neighbourhoods can be defined in an arbitrary manner, but are generally conceived as adjacent cells using a 3 x 3 von Von Neumann neighbourhood or a 3 x 3 Moore neighbourhood. The radii of the neighbourhood can of course be varied, thus incorporating more cells within a cell's neighbourhood. Depending on the amount of cell states, the transition rules and the starting conditions, CA can show very complex forms of behaviour (Wolfram, 2002) mimicking a multitude of behavioural patterns found in real-life situations. In fact, complex behaviour can be achieved already using a very simple rule-set and cell states as demonstrated in the popular 'Game of Life' (Gardner, 1970).

In the early 1980s CA and urban (growth) models merged into what are currently referred to as Geographic Automata Systems (GAS)(Benenson et al, 2004). White (1998) formalized the framework for applying CA as a model for urban representation as follows:

- *Cell Space*. Typically, CA developed for urban modelling are defined on a 2-dimensional lattice. The cell size, depending on the application, ranges typically from 250m down to tens of meters. Cells shapes may be irregular (e.g. representing cadastral units) but are typically regular. Cells typically represent land use or land cover and might incorporate vectors of specific properties (e.g. height, slope).
- *Cell State*. Cells can use binary cells states (e.g. built-up or empty), a discrete set of cell states (e.g. land use or land cover classes) or fixed states. The latter

state can be used to represent e.g. functions that are not open to transition like rivers or parks. The classification by White can be extended by continuous cell states representing a quantitative property (e.g. population size).

- *Transition Rules*. Transition rules can be deterministic or stochastic and ranging from simple to elaborate. White (ibid) does not mention a difference between general rules and so called totalistic rules. The former rule set is based on neighbourhood states using the location of the neighbouring cells while the latter uses aggregate (or mean) values of those neighbouring states regardless of their location. Note that general rules grow exponentially by adding cells states and increasing neighbourhood size, making this method less applicable for GAS. Furthermore, transition rules in GAS are generally fixed, but can be dynamic over time (e.g. Caglioni et al, 2006). Typically, transition rules are used to change land-use states depending on the land-use states of neighbouring cells.
- *Neighbourhood*. Von Neumann and Moore neighbourhood definitions are convenient for many physical processes were field effects are absent and interaction occurs only through contiguity. Yet, in general larger neighbourhoods are often more appropriate within GAS since local decisions are often based on information covering a wider range than proposed by the 3x3 neighbourhood definitions; e.g., urbanization pressure is not only exerted by adjacent plots but also by those within a wider neighbourhood. To extend White's notion of neighbourhood it is important to realize that also global information can be used as input for transition rules. E.g. macro-economic conditions, which clearly have an impact on urban growth, can be applied while stemming from outside the direct neighbourhood boundaries.
- *Time*. Generally, time within GAS is discrete and transition rules are applied in parallel (i.e. cells update their states simultaneously). Yet, some part of the GAS may run at different time scales. E.g., cells subject to seasonal inundation by flooding might operate on a timescale of months while those outside the floodplain could be updated every year. Furthermore, updates could be performed asynchronously using some ordered sequence. As a rule, the timescale on which the GAS operates depends on what process is simulated. Urban growth might be simulated using yearly intervals, while migrating distributions of people caused by commuting might be simulated using time steps of hours.

These elements clearly show that GAS are very closely related to the earlier introduced CA; they are basically nothing more than an instantiation of CA extended with some properties typically connected to geographical models. To equip GAS to mimic actual urban development, the cell space, cell states, transition rules, neighbourhood and time have to be composed in such a way that the model shows the same urban growth patterns as can be perceived from actual data. This introduces a number of sub-problems:

- *Representation*. Since cells basically can represent any property of any size, it is vital to choose cell sizes and states that are expressive enough to provide the requirements initially set to the model. For instance, it is ineffective to use a 1m cell-size when examining the distribution of land-use classes. In practise though, cell sizes are determined by the acquired remote sensing data. For urban growth models, typically resolutions of 30m and lower are used. Regional models though often use resolutions up to 500x500m (White et al, 2000). Apart from the cell's dimension, cell states should be expressive enough to provide the required level of information (e.g. built-up areas or actual land use classes) and use the amount of properties required to define effective transition rules. Thus, cell states could be defined as built-up or empty while they use a multitude of properties, e.g. density, age, slope, proximity to infrastructure, etc.

- *Transition Rules*. Transition rules can be predefined, e.g. a cell's state changes from empty to built-up when 4 or more of its neighbours have the cell state built-up. Typically, urban growth models use predefined rules based on expert knowledge or generalized empirical analysis. Although expert knowledge is a good starting point, chances that the GAS actually confirms observations made from real data are relatively small. Furthermore, since urban growth patterns differ per city, no general rules can be defined that are applicable in every case. Typically, this approach is therefore used to study the outcomes of rule-based scenarios. An alternative is to develop an autocalibration procedure, in which transition rules are adjusted automatically in order to mimic the observed LULC transitions (e.g. Yang et al, 2008; Li et al, 2002). Defining the rules thus becomes an optimization problem.

- *Evaluation*. If urban growth models aim to mimic actual urban development, metrics are required to assess if this is indeed the case. Typical metrics applied for verification of urban growth models are pixel-to-pixel comparisons (e.g. Næsset, 1995) as well as a range of other criteria that attempt to measure if the characteristics of the projected LULC distributions are similar to those observed. These include for instance the fractal dimension of the spatial patterns (e.g. Batty et al, 1994), which expresses the complexity of an observed spatial pattern in a single scalar. All of these metrics have particular strengths and weaknesses. This is why often a combination of metrics is used to express to measure the goodness-of-fit. Nevertheless, human interpretation should not be underestimated. Straatman et al (2004) note that "The eye of the human model developer is an amazingly powerful map comparison tool, which detects easily the similarities and dissimilarities that matter, irrespective of the scale at which they show up".

The aim of mimicking the occurrence of actual (i.e. observed) LULC transitions makes the development of urban growth models a regression problem in which the transition rules (and thus the neighbourhood) become the variables to be optimized. This also implies that transition rules are not necessarily formed by logical clauses (i.e. as if-then-else statements) but can be expressed in terms of weights. Within the domain of machine learning (e.g. Mitchell, 1997) many methods have been developed to attack such classes of problems. In this case application of a supervised learning algorithm (e.g. a neural network, genetic algorithm, Bayesian network) could optimize parameters within the transition rules and neighbourhood settings of the growth model depending based on the outcomes of the applied evaluation function. In this way, the model is using actual data as a training-set to optimize the parameters. Supervised learning is an iterative process which finishes whenever some stopping criterion is met (e.g. a minimum error threshold).

Since CA are decentralized (transitions are based on local boundary conditions for individual cells), transition rules have to be developed that can absorb unique mappings of input values to output values. This implicitly claims that such rules exist. In other words: within actual data, an input pattern at t should always lead to a specific pattern at t+1. This might not necessarily be the case; i.e. similar circumstances in two locations might not lead to the same results. For instance, even though a plot might be profitable to develop from farmland into a residential district, specific circumstances (e.g. social pressure) might prevent such a transition. This is a fundamental problem and has been coped with to some extend by applying so called constrained CA (Engelen et al, 1997). After application of transition rules, the GAS generates a ranked candidate list of cell transitions based on the suitability for transition. Based on a threshold value or on an actual number of transitions, which concurs with the observations, the transitions are executed. Although this method extends growth models with evaluation method of the probability of transition rules, it does not provide a means to catch 'exceptions' in an actual urban growth process. These can only be accommodated by finding the differences in the provided data that discriminate these cases from similar cases.

A further notion is that transition rules might change over time. In majority of CA-based models, the transition rules generally remain stable throughout application of the model at different points in time. In practice though, urban growth isn't necessarily stable; often urban growth is characterized by sudden burst and periods of relative stasis. One way to tackle this is by adding an independent variable that controls the growth rate (e.g. Clarke et al, 1997). Another method is by making the transition rules time-dependent, thus decreasing or increasing their impact to actual transitions over time. More generally though, the phenomenon of urban growth itself is problematic to the extent that urban development does not occur in parallel at discrete time-steps. In other words the notion time, which is often discretized in steps of one year, might need to be adjusted. Instead of parallel transitions, the CA-based model could be equipped with asynchronous transitions (Turner, 1988) by using some ordered sequence. Theoretically, this would change the behaviour of the model considerably because the neighbourhood states of individual cells can differ depending on the ordering sequence at the moment the transition rules are applied; different ordering principles could result in different output patterns. For instance, ordering could be based ranking of candidate transitions in a constrained CA thus implicitly assuming that cells with a high growth potential produce knock-on effects within a wider neighbourhood. This might be a method to confirm the notion that areas with better conditions for urban development produce larger growth than areas with less optimal conditions. Yet, asynchronous updates in urban growth models have received little attention by the research community (Benenson and Torrens, 2004), possibly due to decreasing manageability of the model's behaviour. But more like due to more practical reasons: the absence of time-series data with a temporal resolution of less than a year. While remote sensing data is available for an increasing amount of cities (Benenson and Omer, 2003), historical data covering several decades is often unavailable.

As within any modelling task, the question remains how much information should be included in the model to provide the desired behaviour and precision. To increase precision in urban growth modelling, many developers have added additional properties as within the model's parameter set. Yet, most models incorporate relatively limited set of geographic features to explain urban growth. For instance, the Slope, Land cover, Exclusion, Urban, Transportation and Hillshade (SLEUTH) model developed by Clarke et al (1997) incorporates a variety of physical properties (e.g. slope, proximity to infrastructure), influencing the potential for a cell to be urbanized. The identification of this set is resulting from a different branch of urban growth models, namely those based on Markov models, which are less focused on topological relations that dominate a

CA-based approach. Benenson and Torrens (2004) state that "According to the Markov view, all factors are equal candidates in influencing land-use changes, and those that are ultimately selected should be chosen by statistical analysis." Landis and Zhang's (1998) model for simulating urban growth in California examined the influence of 6 factors on their impact on land-use transitions:

- *Community level:* employment change, household change, total households, employment, job/householder ratio;
- *Accessibility of a cell:* distance to San Francisco, San Jose, freeway interchange, nearest station for the local train, and to areas for which essential urban services are committed;
- *Physical state of the cell*: slope
- *Policy constraints (dependent on community)*: consideration of whether the unit is prime agricultural land;
- *Cell neighbourhood*: fractions of uses — residential, commercial, industrial, public transportation, and vacant — within a 200m radius of the cell;
- *Cell's externalities*: distance to the nearest cell where commercial, industrial, or public used prevail.

An important outcome of this study (ibid) was that the influence of individual factors differed per region; only the distance to freeway interchanges (accessibility of a cell) turned out to be significant for all regions. This introduces a fundamental problem: if urban growth might be determined by an unknown set of drivers, an effective model can only be developed after performing a sensitivity test in which the potential impact of many features is tested on observed LULC transitions. These features are a subset of a potentially much larger dataset that includes physical as well as socioeconomic factors that might explain urban growth at a specific location. Depending on the size of the subset this could significantly increase the data requirements for developing an urban growth model. More importantly though, incorporating a large set of features to explain LULC transitions increases the model's complexity. This can easily lead to uncontrollable behaviour. Although the setup of any CA-based model is relatively simple, because of the vast amount of local interactions between cells and their neighbourhoods, the behaviour from a system's perspective can be complex (e.g. Wolfram, 2002). Subsequent changes in transition rules might not lead to desired outcomes simply because the nonlinear transition functions have become overly complex which comprises control over the model. Unfortunately, the number of systematic studies on the impact of using vast sets of parameters in urban growth models is limited (e.g. Bäck et al, 1996). From a modeller's perspective it seems advisable to use as much information as possible from 'within' the data and only use external parameters when absolutely necessary. As for every model, also here Occam's razor applies.

Related to this issue is how urban growth dictated by a top-down urban planning framework influences the performance of urban growth models. Although maybe counterintuitive, it is easier to accurately predict `organic growth' with CA-based urban growth models than growth based on strict application of land-use plans since the latter might be the result of a top-down and possibly idiosyncratic (political) decision framework. Top-down planning directed urban growth is not necessarily based on locations with the highest suitability; the choice of development locations might be based on for instance a future development vision of an area identified as less suitable for building. Also, changes in regulations might impact the 'suitability landscape' significantly. As a logical consequence, this makes CA-based urban growth models fit for indicating 'unnatural growth' (Benenson and Torrens, p.119, 2004).

The discussion of all these factors that complicate urban growth modelling might give the impression that scientific exploration of this domain provided only fruitless results. This is not the case. Both in theory and practice, urban growth modelling based on a CA-approach has been applied successfully in simulating and predicting urban development for many different case studies (e.g. Moghadam et al, 2013; Shi et al, 2007; Li et al, 2000). Nevertheless, the use of urban growth models is not common practise. This is partly due to a lack of legitimacy in the planning community which was already identified decades by Lee (1994). Yet, these initial reservations should by now be overcome especially since the use of GIS and computer aided design tools has been embraced by the planning community.

## 1.3 Urban growth and scenario development

One of the possible approaches to cope with future uncertainties is to incorporate scenarios. In most cases, scenarios are not necessarily future predictions but potential pathways in which a certain system can develop (e.g. Berkhout et al, 2002). Depending on the system they are describing, scenarios can range from quantitative explorations of key variables to narratives in which storylines are used to illustrate how a system might develop in the future. In adaptive planning frameworks (Walker et al, 2013) scenarios are focussing on possible future conditions in which the impact of a certain policy, strategy or measure is evaluated. Thus, scenarios are mostly used as a technique for assessing the robustness of proposals against for instance the impacts of future climate change. Generally, IPCC scenario families (e.g. Nakicenovic et al, 2000) are used that in some cases are downscaled to develop regional scenarios expressed as alternative intensity-duration-frequency curves for rainfall, water stages or other climate related stresses (e.g. heat or drought-related). These scenarios are continually updated because of observed data, updated model outcomes and new insights. Scenarios for socio economic development include the UN scenarios for population growth and GDP development, which in turn have been 'downscaled' to country-level. In the UK, the Foresight scenarios reach widespread application especially in the development of energy and environmental policy development (e.g. Hall et al, 2005). These explored four different future 'societies', where the relative dependency and state of local and regional communities defined different future narratives (e.g. 'world markets', 'local stewardship'). These scenarios were further developed into quantitative figures about for instance the housing market, energy demand, etc.

Note that there is a fundamental difference between these so-called exploratory scenarios and more aspirational driven normative scenarios, which describe the pathway towards a predefined goal or objective. Normative scenarios are typically used to identify the key ingredients required to achieve the objectives. This application is beyond the scope of this research where scenarios are typically used to explore future outcomes by means of extrapolation.

Although many scenarios are spatially bounded (i.e. they describe possible futures for a specified geographic region), they are not spatially explicit; their variables do not contain spatial units but are expressed in numerical trends (e.g. Semadeni-Davies et al, 2008). Yet, in assessing the impact of phenomena where impact is spatially constrained, i.e. the impact is limited to a specific local area, spatially explicit scenarios can be critical for the development of proper results. In many cases spatial scenarios lead to non-linear outcomes that cannot be determined merely by statistical operations. Essential for spatially explicit scenarios, is the development of mechanisms that drive spatial trends. For instance, a drive towards a compact city dominated by high rise buildings or sprawl-oriented policy will require different mechanisms to achieve such a result. By for instance changing 'push' and 'pull' factors between clusters of built-up areas, different scenario outcomes might be reached.

In many cases though, a straightforward pathway is the development of a business-as-usual scenario; an extrapolation of current trends based on time series analysis. In case of spatial scenarios, the identification of such trends as well as the mechanisms that lead up to them, might not be easily determined. Sophisticated algorithms, like those described in the section on urban growth models (Chapter 3), are required to determine factors, relations and parameterisations of those relations. Furthermore, the issue of determining what is a representative interval from which the time series is used to derive the trends from is open for interpretation. Regime shifts in for instance national politics can have radical implications that might not be representative for longer periods. Yet, the epochs that appear from such historical alterations might require separate treatment. An example is for instance the implosion of the Soviet Union, which in many regards meant a radical trend changes that had a big impact on the development of cities (e.g. more sprawl). In this case it therefore makes sense, to extrapolate data only from source data from the mid-1990s onwards and disregard data prior to that period.

An attempt to integrate both climate change scenarios, socio-economic pathways as well as spatially explicit growth scenarios was made in the Collaborative Research on Flood Resilience in Urban areas (Kurzbach et al, 2013). Yet, the obvious interlinkage between socio-economic pathways and the spatial consequences was only followed for a single case study (Dhaka) and merely conceptualised as an adjustment of future growth rates.

## 1.4 Urban growth and flood risk

### 1.4.1 Cities and riverine flood impacts

Aside from a set of urban growth scenarios, future flood risk assessment obviously requires a set of flood maps, models or a set of expressive indicators from which an evaluation can be made. Within the scope of riverine flooding, the available data is typically scattered over many projects, agencies and scale levels. In Europe, a major driver for a standardisation of flood risk maps is the EU Flood directive 2007/60/EC (European Council, 2007). The directive states that flood risk maps need to be pre-pared for frequent, moderate and extremely rare events with corresponding return periods of 10, 100 and > 100 years respectively. Yet, according to De Moel (2009) still significant differences exist between how the directive is interpreted and applied by the different EU member countries. In many cases, countries limited themselves to the development of flood extent maps which do not include flood depths or actual impact assessment (e.g. vulnerable objects or damages). Furthermore, in some cases the maps are limited to 100 year-events, which therefore do not provide an indication about the proportionality or graduality of flood impacts (De Bruin, 2005) derived from the characteristics of frequent, moderate and rare events.

One of the main issues in urban flood risk assessment is the level of detail that is re-quired to fully express the differentiation between different assets that make up the city (e.g. Veerbeek et al, 2009). Typically, a 1D numerical river model is extended to serve a local case-study (e.g. Horritt et al, 2002); water stages associated to different conditions are translated into overland flow based on a level of detail that is expressive enough to adjust to the local landscape morphology and/or building contours that populate the area of interest. This poses extensive data requirements and is therefore often limited to individual case-studies instead of extensive regions (e.g. Apel et al, 2008). The level-of-detail that is customary for stormwater modelling, which often is limited to drainage units (i.e. neighbourhood scale), is therefore only rarely found for flood assessments of riverine flooding. Nevertheless, new models (e.g Volp et al, 2013), increasing availability of high resolution elevation data (e.g. LIDAR) and low-cost availability of extensive computational resources will make the availability of precise flood maps available for an increasing number of case studies and events. This in turn allows for advanced flood impact assessment on individual building and street level (e.g. Veerbeek et al, 2009).

The development of global models is hampered by extensive data requirements and computational requirements (Woodland et al, 2007). Obviously, correct representation of river sections along extensive stretches as well as standardised design events makes this a complex endeavour. Nevertheless, models have been developed for a European context using a simplified rainfall-runoff model (Roo, 2000) and a subsequent analysis by Lugeri et al (2006) for 13 European countries. This model uses a 1 km grid and a classification of 5 qualitative hazard levels ('very low' to 'very high') instead of return periods. Land use was represented by using the CORINE 2000 land cover classification which is based on 250m grid cells. The assessment did not include future projections, including climate change scenarios or land use changes. In a European context, the latter remains relatively stable since urban development is modest compared to for instance Asia and Africa. For a uniform coverage of areas in those regions (including cities), only flood maps produced by global models are available. These have been developed for instance by Herold et al (2011), albeit for floods with a 100Y return period. These further evolved into models developed by Papenheim et al (2012) and Winsemius (2012) who included a whole range of return periods, between 2 and 500Y events and 1 and 1000Y events respectively. Both these models are based on resolutions of 30 arcsecs which corresponds to around 1km at the equator.

The pan-European and global models all focus on flood hazard, i.e. they provide the flood extent or flood depth distribution but do not include an assessment of the impacts on rural as well as on urban areas. Large scale flood impact assessments that include riverine flooding have been performed by Jongman et al (2012), who have focussed on the combined impacts of coastal and riverine flooding with a global coverage. While the assessment was limited to 100Y events and based on a rather coarse resolution of 30 arcsecs, the assessment included future land use changes with a long term horizon until 2050. Yet, the land use changes were not based on actual geographic modelling, but depended on statistical extrapolations using for instance World Bank population and GDP estimations. Thus the monetary value or population of single landuse cells was changed by multiplication based on the statistical projections instead of integrating actual transitions of landuse cells. The outcomes were presented as aggregates at subcontinental scale, which obviously differs from the scope of this study. Nevertheless, the outcomes showed a significant increase of flood impacts from both contributors. Especially, the projected impacts for Asia increase almost 4-fold.

## 1.4.2 From imperviousness to surface runoff

One of the most significant effects urbanisation has on the water system is associated to the increase of impervious surface covers, i.e. materials covering the ground that prevent infiltration of water into the soil (Arnold and Gibbons, 1996). The construction of buildings, roads and parking lots typically lead to soil sealing of extensive areas, which in turn results in substantial additional surface runoff. Apart from these unambiguous examples, other man-made as well as natural surfaces can be so heavily compacted or saturated as to be functionally impervious (e.g., bedrock outcrops, dirt roads). When used as a landscape indicator, impervious surface cover is typically indicated as a percentage of the land that is covered with impervious materials; the impervious surface ratio (ISR).

Establishing a relation between urbanization and the increase of problematic levels of surface runoff (i.e. floods) has been the topic of extensive studies. Transformation of land into built-up areas leads to larger and more frequent floods (e.g. Leopold, 1968) expressed as changes in peak discharge, lag time, flood frequency and total runoff. Peak discharges can increase two to four times as a result of urban development, lag times decrease correspondingly to one and a half to one-fifth (Chin, 2007). These effects have been shown across various scale levels. E.g., Perry and Nawaz (2008) show how hard surfacing of domestic gardens in suburban areas leads to increasing susceptibility towards urban flooding in Leeds, UK over a 33 year period. Perry and Nawaz estimate a 12% increase of annual runoff caused by this very low-level phenomenon. Most studies though are based on catchment scale showing how urbanization of some regions in the catchment affects discharge distribution and level (e.g. Brun and Band, 1999). Typically, the studies observe increases in streamflow hydrographs resulting from increasing ISRs caused by urbanization, thus increasing the water yield.

Although impervious surface maps can be derived from high resolution satellite photographs, the relatively limited footprint and price of such images often imposes difficulty at the moment data on larger areas is required (Canters et al, 2006). More efficient ways of generating impervious surface maps are using medium-resolution images (e.g. Landsat ETM+, Aster, etc.) and try to downscale these to higher resolutions. Typically, this process compromises accuracy. In order to overcome such problems, sub-pixel regression and sub-pixel classification methods have been developed. These rely on associating medium-resolution pixels to representative high resolution tiles from which the impervious surface ratio (ISR) is derived. Alternatively, impervious surface maps can be obtained by simply applying pre-defined ISR ratios associated to LULC classes or derive these from existing LULC maps. As opposed to the often raw data from satellite imagery, these maps are generally ready-for-use, geo-referenced and checked by (governmental) institutions that publish the data, which in most cases ensures accuracy. Typically, urbanized areas can show a wide range of ISRs ranging from about 20% for low density built-up areas to close to a 100% (e.g. SCS, 1985), although typically mean values are used which obviously introduces errors (Tan, 2008).

Runoff coefficients, expressed as the ratio of runoff to precipitation (e.g. Bedient and Huber, 2002), are closely related to ISRs (Canters et al, 2006). For general assessments these can be estimated by using a land-cover classification in a similar fashion as for estimating imperviousness. If an impervious surface map is available, runoff coefficients can also be derived by applying a function to the imperviousness levels, as part of Rational Method (e.g. Singh, 1992) to calculate urban stormwater loads.

Shi et al (2007) showed that for the important economical region of Shenzhen, China, runoff coefficients increased dramatically because of the rapid urbanization process taking place over the last decades. Within the Buji River basin, located in Shenzhen, urbanization level increased from approximately 2% in 1980 to 59% in 2000, increasing runoff coefficients by 32% during dry conditions for a 90% storm probability. Large scale comparative studies for different urban case-study areas in relation to runoff increase are rare. Chin (2007) studied the urban transformation of river landscapes comparing research results from over a 100 studies. Especially since runoff was only a subtopic in this study, the study offers merely an overview of research papers than providing a thorough quantification of runoff increase in urban areas worldwide. Specific research on the influence of future urban growth scenarios on runoff generation has been performed by Choi and Deal, 2007. For the Kishwaukee River basin, USA, they developed an urban growth model based on different economic development scenarios to assess hydrological impacts of urban growth for the year 2051. What is interesting from this study is that annual runoff only increases by 1.7% in 2051 using the most favourable economic development scenario while built-up land cover increases by 6%.

To actually estimate changes in runoff caused by runoff coefficient variations, methods have been developed which usually apply hydrographs that show runoff over time for a given catchment. The crest of the hydrograph defines the peak flow which is usually determined synthetically since actual measurements of hydrographs (continuous measurement of discharges) are often not available. Peak flow can be determined either through deterministic models or by regression (e.g. Savic et al, 1999). Deterministic models are generally based on the Rational Method (Kuichling, 1889) and the Soil Conservation Service (SCS)(SCS, 1957) runoff methodology. Both methods are simplified procedures, the methods are considered sufficiently accurate for runoff estimation when only general approximations are needed. It is important to realize that the Rational Method as well as the SCS model ultimately provides numerical output on runoff volumes. For obtaining insight in the spatial distribution of surface runoff (i.e. the overland flow), more intricate models have to be used based on Saint-Venant equations for two-dimensional shallow water flow. Typically, such models require a mesh representation of area features for solving the equations for adjacent cells using a finite element technique. The so called "roughness coefficients" are used to express physical characteristics resisting water flow (e.g. Horritt et al, 2001) in the cells. Yet, in an urban setting the physical characteristics of the environment are primarily represented by a meshed representation of a digital elevation model. In order to correctly simulate the propagation of overland flow, such a model requires a high level of detail to represent actual urban features like street profiles, buildings, local depressions, etc. This in turn requires considerable computational resources in order to represent large urban areas as well as to solve the equations required for simulating the flood. Furthermore, this also sets requirements for prospective simulations in settings where urban growth is projected: the representation of future features (e.g. new residential areas) should contain a detailed design which is only possible in case an actual urban design for the area in question already exist.

## 1.4.3 Towards the drained city

The urban drainage in cities is obviously not only dependent on the infiltration characteristics of the territory's topsoil and subsoil. In virtually all cities some kind of engineered drainage system has been constructed to convey stormwater into receiving waterbodies (i.e. a lakes, river or directly into the sea). The level of development as well as the integration of the drainage system into the complete scope of water services that the city provides, is often a result of a historical evolution in which the provision of drainage can be considered a logical step following the provision of i) water supply and ii) sewage as basic building blocks (Brown et al, 2009). Obviously, this classification is a merely a crude labelling that ignores the extensive differentiation in the development of water-related services within cities. In urban drainage, the capacity of the local drainage is often closely connected to the progression of urban development over time. Historic downtown areas typically face drainage issues due to lower design standards, while new residential areas at the fringe of cities often incorporate stormwater management systems based on updated requirements. Furthermore, gradual densification of downtown areas often increases the pressure on the already aging drainage infrastructure. Especially in cities where public spending on infrastructure is limited, this often leads to extensive waterlogging during even moderate rainfall events that are well below the initial design event. The effects are aggravated in combined sewer systems where surcharges often lead to significant health and environmental impacts (e.g. Mark et al, 2015) .

Urban drainage networks are typically modelled as a combination of a surface/subsurface network of one-dimensional open channels (i.e. surface flow paths) and drainage pipes. These so-called 1D/1D coupled models are well established across many cities to simulate drainage performance and estimate flood volumes for a given rainfall event. Typically, the flow paths in such models are developed from the street network which in especially high density urban settings represents actual conditions. Yet, in most cases the modelling of surface flow is preferably performed by a 2D model which is combined with the 1D subsurface model. The main advantage of 1D models are their relative modest computational requirements, which means that flow in vast networks consisting of large numbers of nodes can be calculated with modest computation resources. A major obstacle for effective use of both model types is the availability of extensive field data needed for calibration in order to mimic actual behaviour. Unfortunately, most of the models are calibrated based on data of events with little or no flooding which often means that the speed and direct runoff volume is often underestimated (e.g. Djordjevic, 2011). In order to match flood extent and depth, extensive data from flood observation is required.

Again, the use of such models in the context of future urban growth projections is problematic since no open channels and pipe drainage networks can be defined when growth is merely expressed as LULC transition of a raster cell. A promising alternative might come from an algorithmic approach where artificial sewer networks are generated using location specific properties (Ghosh et al, 2006). Yet, research in this area is very limited. More generally though, there might be a considerable need for relatively crude yet rapid modelling approaches that can cover extensive areas. This might not only be profitable for estimating future urban flood issues in rapidly developing cities but is also essential in the current context of many cities in the developing world. Here, data is often sparse and conditions are volatile which severely limits the possibilities for the development of extensively calibrated models. Furthermore, many of these cities are located in regions with a monsoon driven climate where especially insight in the consequences of peak rainfall events is required. Relatively crude approximations are therefore often sufficient to identify flooding "hotspots".

## 1.5 Flood vulnerability assessment

The science of vulnerability assessment is characterized by the development of a great number of conceptual frameworks within many knowledge domains (e.g. Turner et al, 2003). Although these frameworks identify large numbers of indicators and therefore implicitly proof the complex nature of vulnerability assessment, it is important to identify communalities. Luers et al (2003) defines vulnerability as a function of the sensitivity of a system to stress and the state of the system relative to a threshold. The sensitivity is defined as the derivative of some stressor function and defines the progression rate of the stress. The state of the system relative to a threshold on the other hand relates to the level of stress a system endures at some moment. Stress is used as a container term expressing a particular hazard a system is exposed to. This leads to the somewhat more accessible risk frameworks, where risk is defined as a function of hazard, exposure and vulnerability (also often replaced by 'sensitivity') (e.g. Luers, 2005). In flood risk assessment the hazard is typically defined as a particular flood event associated to an estimated return period or frequency, the exposure as the area-of-interest covered by the flood (e.g. a neighbourhood) and the vulnerability as the actual impact the area-of-interest suffers. In an urban setting, expressing the vulnerability is often problematic due to its inherent complexity. Impacts can be tangible, intangible, direct as well as indirect (e.g. economic impacts due to interruptions in a supply chain). Many of the impacts can be expressed in monetary values, but are difficult to estimate due to the number of components that are affected as well as because of the differentiation in components. In flood risk assessment, direct damages are calculated by stage-damage functions that associate inundation depths to monetary losses since the affected components need to be repaired or replaced. Compared to the differentiation of exposed assets (e.g. different housing types) in flood prone areas, the number of available stage-damage curves that describe their vulnerability is marginal. This renders especially ex-ante flood damage assessments to mere approximations.

Typically, flood impacts are expressed as those related to a flood with a specific return period (e.g. 100Y event) or as the weighted mean of the expected damages from a range of events: the estimated annual damages. Yet, both methods do express very little about the impact progression for decreasing likelihoods of occurrence. In an attempt to better express the concept of flood resiliency (e.g. Gersonius, 2012), De Bruin

(2004) applied the concept of graduality to the domain of riverine flooding. Graduality determines to what extent increased level of stress on a system results in a proportional response. Systems showing low degrees of graduality can show disproportionate reactions when some threshold value is reached. This identifies if systems show some degree of self-organized criticality (e.g. Slanina, 2005), i.e. a stable state to which the system can "bounce back" when coping with a range of stressor levels ,accompanied by a sudden (and often irreversible) state change when a specific stressor level (i.e. threshold value) is reached. Insight in the graduality might be important to identify if a system (e.g. a city) might endure irrecoverable impacts once the flood hazard exceeds a threshold value (e.g. Mens et al, 2011).

## 1.6 Converging to the scope

The previous paragraphs have introduced the issue of urban growth and its potential impacts on future flood risk with a particular focus on riverine and pluvial flooding. To actually assess that impact of urban growth, a number of requirements and knowledge gaps have been identified. Most notably, the absence of spatially explicit urban growth scenarios for fast growing megacities and the challenges in assessing different types of flood risk from those scenarios were considered important omissions in current research. In the following chapter, the resulting requirements and gaps will be more formally described in a set of research questions, hypotheses as well as the proposed methodology to address them. From this chapter the initial conclusion is that although the problem seems extremely urgent and should be a priority in the domain of flood risk assessment, studies that attempt to address the issue in a comprehensive comparison are practically absent.

It is safe to assume that not all issues that have been raised will be solved in a comprehensive and optimal manner. Nevertheless, the big advantage is that even if some of the outcomes are suboptimal, they are still able to fill a gap that is currently still mostly "uncharted territory". The contribution made by this study might prove to be a first step towards the application of LULC models in the domain of climate adaptation.

## 1.7 Reader's guide

Apart from the introduction covered in this chapter, this study is subdivided into 8 additional chapters:

- *Research Questions and Methodology.* The state-of-the-art and the presented research gaps from the introduction are further developed into primary research question, a set of sub-questions as well as a series of hypotheses to test the validity of possible outcomes. Furthermore, the chapter introduces the requirements for the methods, models and evaluation in order to answer the research questions.

- *Memetic algorithm optimised urban growth model.* In this chapter the urban growth model is presented. Focussing on the case study of Beijing, the approach, technical setup as well as an extensive coverage of the quality of the outcomes is provided.

- *Growth projections.* The growth scenario in combination with the urban growth model is applied to a selected set of fast growing metropolitan areas. This chapter discusses the resulting outcomes for the different cases, focussing on trends and differences. Special emphasis is put on the analysis of the growth characteristics and the drivers and constraints that might influence growth.

- *Future riverine flooding in megacities.* The growth projections are combined with sets of flood hazard scenarios to estimate how future flood risk develops. The chapter covers different aspects of flood risk analysis, including the translation into operational metrics that are applied to the selected case studies.

- *Assessing the effects of urban growth on urban drainage.* This chapter is similar to the previous chapter, but primarily focusses on pluvial flooding issues. Also here, a major part of the chapter covers the question how to develop and apply operational metrics given the constraints resulting from the growth projections.

- *Adding depth: Estimating flood damages in Dhaka.* Given the large number of case studies and extensive areas, covered by many megacities, a case study further developed in-depth. Focussing on the flood prone city of Dhaka, a flood damage assessment is presented in which urban growth is used as the main driver.

- *Further explorations.* The application of urban growth projections is obviously not limited to flood risk assessment. In this chapter, two alternative applications are presented focussing on water quality in slums and on precipitation changes due to urban growth-induced development of the urban heat island.

- *Towards an argument.* Finally, in this last chapter the outcomes of the different chapters are summarized and related to the initial research questions. The hypotheses are evaluated and the study is finalised with a conclusion, discussion and future outlook on further research.

## 1.8 Embedded research projects

Most of the outcomes in this research are based or directly drawn from work in previous projects. An important pillar for the development of the growth model was the work performed in the project: Collaborative Research on Flood Resilience in Urban areas (CORFU) (Djordjević et al, 2011). Within the scope of this project, the model was developed and tested in the case study areas of Beijing, Dhaka, Mumbai and Seoul. The LULC classification for these cities was also performed within the scope of the project and was in the cases of Beijing and Dhaka validated using datasets provided by the municipal development corporations. Chapter 3, which covers the setup of the urban growth model, was previously published in Computers, Environment and Urban Systems (Veerbeek et al, 2015). Further application of the model in the other case study areas was performed outside an actual project context.

Activities required for the riverine flooding assessment were performed within the context of the Cities, Water and Governance-project, commissioned by the PBL Netherlands Environmental Assessment agency (Veerbeek et al, 2016). The main body of this report is integrated in chapter 4 on riverine flooding.

The activities in the context of pluvial flooding were performed exclusively for this study and were not part of an actual project. The peripheral studies on urban heat island-induced local precipitation changes as well as the section on future pollution loads due to slum development (Section 8.1), were based on two MSc theses.

A more extensive coverage of the sources of the individual chapters, including the outcomes is provided at the start of each chapter. In cases where work draws on results from third-party contributors (i.e. other authors), the sections as well as the respective authors will be indicated.

# 2. Research Questions and Methodology

## 2.1 From state-of-the-art to research questions

The previous chapter introduced the issue of rapid urban growth, the possible impact on future flood risk and emphasised the underexposure of urban growth in future flood risk assessment. In summary, the basic argumentation follows these six lines:

1. Apart from climate change, future flood risk is significantly affected by urban growth which is progressing at an unprecedented rate;
2. Urbanising deltas as well as upstream regions along major rivers cause extensive biophysical changes that significantly alter the hydrological characteristics;
3. Application of explicit urban growth scenarios is required to assess future changes in risks of natural hazards, and particularly of floods.
4. Urban growth should be a standard component of future flood risk assessment as well as for the development and evaluation of future adaptation strategies.
5. The outcomes provide insight how cities perform in relation to different types of flooding: coastal, riverine and pluvial. Possibly, some cities show similar trends.
6. A comparative study is required based on a consistent and uniform approach which sets specific constraints and requirements for the methodology, datasets, models and analysis of outcomes.

Given that for the assessment of future growth in induced coastal flood risk a considerable amount of pioneering work has been performed (e.g. Hallegatte et al, 2013; Nicholls et al, 2008) the focus of this research is therefore on riverine and pluvial flooding, for which research is still limited. The argumentation in combination with the identified knowledge gaps supports the following overarching research question:

*What is the comparative impact of future urban growth on the development of riverine and pluvial flood risk of fast growing metropolitan areas?*

The objective of this study becomes to (i) develop a future urban growth scenario for a set of rapidly growing large cities (ii) to integrate the outcomes with flood data for riverine and pluvial flooding and (iii) to evaluate the characteristics of future flood risk trends as a function of the growth scenario. This leads to a number of requirements:

- The development of a spatially explicit urban growth model;
- A sizable set of fast growing large metropolitan areas;
- An appropriate method to estimate riverine and pluvial flood risk;
- A method to compare the outcomes across the different case studies;
- A characterisation of the relative changes in flood risk; normalised in relation

to the projected urban growth;

Combining these objectives urges for a refinement of the main research question into a series of sub-questions:

RQ1. How can a context specific, spatially explicit urban growth scenario be developed that is based on a robust extrapolation of past trends, comparable data sources and with a relatively high level of detail and precision?

RQ2. How can urban growth affected flood risk be estimated and compared given the considerable size as well as the differentiation within and across the metropolitan areas?

RQ3. What are the spatial characteristics of the growth projections and subsequent flood risk for the selected areas?

RQ4. In which cities does normalised flood risk increases at a disproportionate rate and do those cities confirm existing expectations?

To answer these research questions a set of criteria and constraints have to be introduced to further narrow down the scope of this research in order to keep the required activities manageable and to converge to a set of outcomes that are robust and accurate.

First of all it is important to recognize that the focus of this study is application to actual cities. The general implications of urban growth on flood risk are well understood from a theoretical perspective, yet a large scale comparison based on projections for actual metropolitan areas has not been performed. Furthermore, the aim for spatially explicit growth scenarios introduces a heavy dependency on spatial data which might be challenging to acquire, especially since for some of the data multiple instances are required at different points in time (e.g. historic LULC maps). Finally, the assessment of the proportionality of the estimated changes in flood risk might not necessarily be straightforward, especially in case proxy indicators are required when flood risk cannot be estimated directly due to gaps in available data.

## 2.2 Hypotheses

To better answer the research questions, a set of five hypotheses have been defined that support claims in the areas of urban growth, riverine flooding and pluvial flooding. These topics cover the body of the presented research. All hypotheses have been extended by a null hypothesis which disproves a particular claim. While some of the hypotheses might seem somewhat self-evident, they are required to develop a complete argument from which the main research question and sub-questions can be answered.

**Hypothesis 1: urban growth differentiation based on a uniform approach**

This hypothesis supports the claim that the spatial characteristics of urban growth differ for every city. A model used for the development of spatially explicit urban growth scenarios, should therefore be calibrated or optimised to mimic the peculiarities of the urban growth patterns for each consecutive case. Yet, given the need for a uniform approach, each case study-adjusted model should be based on a similar generic model. So, the hypothesis should ensure that two similar yet differently calibrated models should produce different growth patterns from the same initial dataset:

$H_1$: *Two identical LULC change-models with different parameterizations are able to produce different LULC patterns from a single LULC distributions used as basemap data.*

$H_0$: *Two identical LULC change-models with different parameterizations produce the same LULC patterns from a single LULC distributions used as basemap data.*

**Hypothesis 2: accuracy of the urban growth model**

The second hypothesis focuses on successfully mimicking observed spatial trends in order to create extrapolations of those trends as future baseline scenarios. This means that the model should be able to reproduce comparable LULC-transitions to those actually observed in the selected case study areas; the model should be able to 'predict the past'.

$H_1$: *The developed LULC change-model is able to produce LULC distributions that correspond to observed LULC distributions using evaluation criteria generally accepted in the current state-of-the-art in LULC-change modelling.*

$H_0$: *The developed LULC change-model is not able to produce LULC distributions that correspond to observed LULC distributions using evaluation criteria generally accepted in the current state-of-the-art in LULC-change modelling.*

**Hypothesis 3: flood risk differentiation**

The third hypothesis addresses the issue that the changes in future flood risk as a consequence of urban growth differ over time and space within and between cities. This means the contribution of urban growth to future flood risk needs to be isolated and that the outcomes should be expressive enough, to reflect spatiotemporal differences. For instance, flood risk should therefore not be calculated and expressed as a singular value (e.g. the area's mean annual damage for the future interval).

$H_1$: *Different spatiotemporal urban growth characteristics within the same flood extent of a given metropolitan area, will result in different spatiotemporal flood risk characteristics.*

$H_0$: *Different spatiotemporal urban growth characteristics within the same flood extent of a given metropolitan area, can result in the same spatiotemporal flood risk characteristics.*

## Hypothesis 4: disproportionate increase of flood risk

Extending hypothesis 3, a major assumption of this study is that the projected growth of some cities produces more flood risk compared to their overall growth would; there is a disproportionate increase of future flood risk. Due to the different size and initial flood characteristics, the future changes in flood risk need to be normalised for a fair comparison. This is captured in the following hypothesis

$H_1$: *The normalised flood risk of two cities with similar growth rates, can lead to a significantly higher normalised future flood risk in one city over the other.*

$H_0$: *The normalised flood risk of two cities with similar growth rates, leads to a similar normalised future flood risk for both cities.*

## Hypothesis 5: alternative results

The final hypothesis supports the claim that the outcomes of this study might lead to an alternative ranking of cities, which supports an alternative prioritisation of urban flood risk management in cities that currently might be underexposed.

$H_1$: *There are cities in this study, where the projected urban growth over the interval 2010-2060 within the flood extent associated to a flood event is higher than outside the flood extent for that same interval.*

$H_0$: *All cities grow equal or less in the interval 2010-2060 within the flood extent associated to a flood event.*

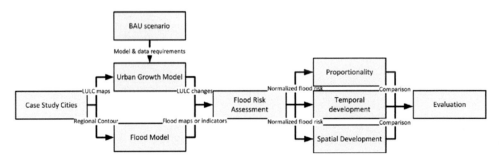

**Figure 1:** Different components of this study and their interactions

The hypotheses provide the logical building blocks from which a final argument can be constructed. Hypothesis 1 and 2 support RQ1: they ensure that the applied urban growth model is specific to each city but is yet uniform. Hypothesis 3 which should ensure different growth behaviour of cities over space and time supports RQ3. Hypothesis 4 is used to ensure that future flood risk can be assessed (RQ2) and tests the evidence of disproportionate increase of future flood risk due to growth differentiation, which is the issue of RQ4. Finally, Hypothesis 5 attempts to test if the outcomes of this study indeed differ from previously produced rankings for coastal flooding. This is covered in the second leg of RQ4.

The different components required for this study follow the flowchart illustrated in Figure 1. The scheme also illustrates the methodology and required major activities required to answer the research questions. In the coverage of the state-of-the art as well as in the main research question, there is a clear distinction between riverine and pluvial flooding. While the urban growth projections and resulting LULC maps, might be used for both flood domains, the methods, models and data to produce flood maps for riverine and pluvial flooding differ. This implies that some of the steps in the workflow, will be performed separately for respective flood risk assessments.

## 2.3 Research Methodologies and skills

Many of the activities in this research are data and model-oriented. Especially the development of the spatially explicit urban growth scenarios requires extensive data sets as well as a model to develop the LULC transitions. To exploit the datasets to their fullest potential, a chosen 'off-the-shelve' LULC model most probably needs to be extended or developed. The development of scenarios for a wide range of case study areas, for which pre-configured datasets and models are likely not available, requires extensive GIS-related activities: LULC classification from remote sensing data, extensive geoprocessing and most of all a good overview of available data resources. The calibration of the LULC modelling requires optimization (i.e. supervised machine learning) which is tailored to the specific constraints provided by the available datasets. The integration of existing flood maps, possibly at a different scale compared to the growth scenarios, requires extensive validation to ensure robust outcomes. Given the extensive volume data created within the research, skills in calculating key statistics that are both concise and expressive is required. Although much of the theory has already been covered in the review of the state-of-the-art in chapter 1, additional literature will be consulted on additional modelling issues as well as background material covering the case studies.

As often, the robustness of the claims in this study (i.e. their applicability in a broad set of cases), is very dependent on the number of case studies. Furthermore, the core of the project consists of a comparative study which makes inclusion of a large number of case study areas only more prudent. However, the analysis of the state-of-the-art already (see Chapter 1) showed that typically, the application of urban growth models is limited to single case studies (i.e. cities). A possible reason for this is that it simply takes too much resources to develop scenarios for a large number of cities. This makes the required volume of activities for this study all the more ambitious.

## 2.4 Methodological considerations

The following paragraphs describe some of the methodological considerations and proposals for some of the tools used in the research. Yet, a more extensive coverage is provided in the respective chapters covering the development of the urban growth model, its application in riverine and pluvial flooding as well as for the additional applications in for instance urban heat island driven local rainfall and water quality issues.

### 2.4.1 Urban growth model and scenarios

In many research projects that involve the development spatially explicit urban growth scenarios, LULC change models are developed from scratch. This resulted in an extensive 'ecosystem' of models often developed for application in single case studies, thus potentially limiting the applicability and usability of models (Benenson et al, 2004). This fragmentation limits the impact of models as well as refinement of existing tools into more sophisticated models that outperform previous versions. The approach in this research project is therefore to use an 'off-the-shelve' model with a proven track record and to adapt and extend the model to adequately address the challenges provided by the included case study areas and to improve the model performance by developing new modules or improve existing ones.

Due to the extensive number of case studies, which are required to support the goal of a comparative study, the model needs to be flexible enough to cope with especially quantitative differences. Since some of the case studies consist of extensive areas, the objective to create comparable results might result in raster maps that comprise of tens millions of cells. This leads to significant computational loads especially during the calibration phase. Producing comparable results also has consequences for the uniformity of the modelling procedure: the data used for calibrating the model should be drawn from comparable datasets. This means that for some of the case studies, data might be omitted that might improve the model's accuracy. This might seem counterintuitive, but since compatibility is a top priority in this research, this choice is defensible.

The growth projections are aimed at a time horizon of about 50 years. Yet, what is an important requirement that the model is able to produce smaller increments (e.g. 5 years) to be able analyse behaviour towards the 50 year horizon. This aim also sets requirements for the range of the historical data used in the calibration.

## 2.4.2 Riverine flooding

While the development of spatially explicit urban growth scenarios for a substantial number of case studies is a formidable task, the modelling of the water stages in adjacent rivers and the subsequent flood events in case of overtopping of river banks possibly requires even more resources. It seems therefore preferable to use existing datasets (i.e. flood maps) that fulfil the requirements as much a possible. That means that they need global coverage in order to accommodate application to case studies located anywhere in the world. Furthermore, they require a level of detail that is sufficient to express the differentiation found at urban scale. Except for precision, the maps are also required to be sufficiently expressive: they need to include flood depths and should cover a range of associated return periods that includes both frequent and rare events.

Determining the different classes of urbanised areas that are intersecting the flood extent provided by the flood maps is relatively straightforward. Still, estimating the resulting impacts requires extensive data about the composition of the urban areas as well as the relation between inundation depths and consequences (e.g. expressed in stage damage-curves). This seems unfeasible for all case studies but might be performed in a few cases as a proof of concept. Still, as in the majority of flood risk estimations, a full assessment including indirect damages as well intangible damages is unlikely not only due to practical reasons, but also due to conceptual obstacles. If future impacts should be estimated in an environment where the city is perceived as a dynamic, ever evolving context, it would be counterintuitive to assume that the vulnerability of areas or particular urban assets classes would remain static. This would require projections about regional GDP-growth as well as other assumptions (e.g. future protection standards) on which the subsequent impact levels on.

A major outcome that might not provide new insights, but is important in terms of 'marketing' the outcomes is the development of a new rank list. Such a list can extend existing lists like those developed by for instance Hallegatte et al (2013) and act as a signpost for the project.

The assessment of future flood exposure of urban areas might possibly involve processing many instances of flood-LULC map combinations. This requires automation to keep the workload manageable and to limit errors. Also, summarizing the outcomes with descriptive statistics requires some automated scripting. In principles this task can be performed in any scriptable geoprocessing environment. Typical tools to perform this task is a combination of Esri's ArcGIS (or the free available QGIS) and Python.

## 2.4.3 Pluvial flooding

As for fluvial (i.e. riverine) flooding it does not' seem feasible to actually cover all case studies with models to simulate pluvial flooding for different local rainfall events. Apart from conceptual obstacles for developing a 1d underground drainage in a context of future urban development, the possibilities to apply an overland flow model are also limited. A major hurdle would be to overcome the relative coarse resolution of the digital terrain maps that are required to identify local depressions for flow accumulation. Current Aster GDEM or SRTM2-data is available at a raster resolution of about 25m, which seems insufficient to effectively identify areas where water accumulates. Since many of the potential candidate case study areas are located on relatively flat alluvial plains, a higher level of detail is required to identify the actual locations where floodwater accumulates. Nevertheless, this might be a possible pathway to evaluate.

An alternative is to focus on proxy indicators like the ISR distribution. For large scale assessments, such an approach is widely used and might therefore be the most feasible method within the scope of this study. Furthermore, additional proxy indicators might be developed that provide more information about the drainage performance of areas within the case studies.

## 2.4.4 Pre- and Post-processing

LULC classification from remote sensing data is an essential step in the development of base maps that are part of the datasets used for calibration the urban growth model. This is essentially a supervised learning tasks, where the errors in a classification are

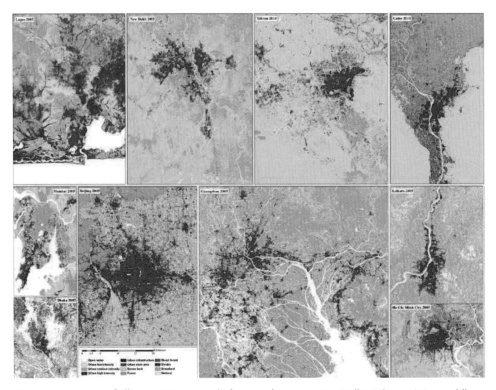

**Figure 2:** Overview of all cities at same scale (top row): Lagos, New Delhi, Tehran, Cairo and (bottom row) Mumbai, Dhaka, Beijing, Guangzhou-Shenzhen, Kolkata, Ho Chi Minh City.

minimised using a training set of LULC classes. The methods are based on a commonly used maximum likelihood classification (ERDAS, 1999) although many more sophisticated methods are available (e.g. Otukei et al, 2010). Although many of the methods have specific pitfalls and advantages, the most important factor in correct classification is the availability of good quality remote sensing data with little or no cloud coverage. Furthermore, in almost all cases ample time needs to be reserved for manual correction of classification mistakes. Although the activities are not always rewarding, putting enough effort in developing high quality base maps is essential for the consistency and reliability of the produced extrapolations (i.e. the growth scenarios). Additional pre-processing task related to the fabrication of base maps require extensive geoprocessing to develop for instance slope-weighted distance maps (e.g. from roads, waterways).

Post-processing will include urban landscape analysis developed by Parent (2009), which is a Python-based script running on ArcGIS platform. Yet the script can be easily adapted to run outside of ArcGIS. Additional spatial metrics might be calculated within FRAGSTATS (McGarigal et al, 1995), which is designed to compute a wide variety of

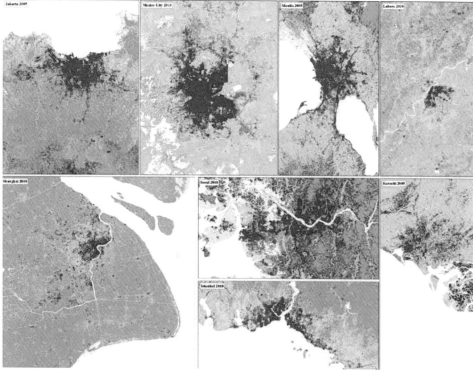

**Figure 2 (cont.)**: Overview of all cities at same scale (top row): Jakarta, Mexico City, Manila, Lahore and (bottom row) Shanghai, Seoul, Istanbul and Karachi.

landscape metrics for categorical map patterns. Also FRAGSTATS can be used within a scripting environment and can therefore be automated for extensive numbers of runs. The emphasis on land cover instead of land use limits application of the data in comprehensive approach to flood impact assessments and purely focus on the ground cover of relatively coarse LULC classes. Obvious differences between cities in neighbourhood typologies, building types and associated densities are omitted in favour of a consistent and uniform approach that can be applied to all cities in the assessment.

## 2.5 Peripheral topics

The application of spatially explicit urban growth scenarios in environmental risk and urban flood management can be used in a variety of domains. Apart from coastal, riverine and pluvial flooding additional research topics include the effects of urban growth on the urban heat island, which in turn changes local rainfall characteristics and thus operates on the hazard component of pluvial flood risk. Another issue which is especially prudent in the developing world is the impact on urban growth on water quality. Dumping of raw sewage and solid waste in streams results in major impacts further downstream into rivers and lagoons and coastal areas. Anticipating future changes in the pollution loads in the stream network might be essential to prevent severe health impacts and ecological catastrophes.

## 2.6 Originality, innovation and potential impact

The literature review in the introduction provides an overview of the state-of-the-art in many of the knowledge domains associated to the issue of rapid urban development and flood risk, including specific knowledge gaps and omissions in for instance model coverage, application or analysis. The individual chapters in this study include a more explicit description of particular knowledge gaps and subsequent methods, outcomes and conclusions to tackle them. Nevertheless, it might be valuable to provide a concise overview of the main contributions of this study.

## 2.7 Selection of case studies

For this study, fast growing megacities (i.e. >10 million inhabitants) have been defined as those with an annual population growth rate of 2% or higher, based on the reference year 2010 (Demographia, 2010). Note that for most cities, the growth rate decreases over time since a stable rate of for instance 2%, would lead to exponential growth. This has resulted in the following set:

- India: Calcutta (Kolkata), Mumbai, New Delhi;
- China: Beijing, Guangzhou, Shenzhen, Shanghai
- Pakistan: Karachi, Lahore
- Indonesia: Jakarta
- Philippines: Manila;
- Korea: Seoul;
- Bangladesh: Dhaka;
- Vietnam: Ho Chi Minh City;
- Egypt: Cairo
- Mexico: Mexico City
- Iran: Tehran
- Turkey: Istanbul
- Nigeria: Lagos

If this list would be updated for the current year, five new fast growing megacities in China would have been added: Chengdu, Harbin, Hongqing, Tianjin, Wuhan as well as Congo's capital Kinshasa. Furthermore, Mexico City and Seoul would have been taken off the list since their annual growth rate is currently below the 2% mark. Furthermore, in the reference year, Ho Chi Minh City did not reach a population of 10 million inhabitants and could therefore not be characterised as a megacity. Yet, the city was included due to the attention the city received from the flood risk community. Currently the city has reached megacity status and is still experiencing annual growth rates above 2% (Demographia, 2016). An illustration of the selection of megacities is provided in Figure 1, where the cities are represented at similar scale. This immediately draws attention to the different dimensions of the urban extent, ranging from about 240 km2 for Dhaka to  about 5800 km2 for the Guangzhou-Shenzhen area. Base statistics of each individual city are provided in Appendix A3.

Apart from coastal flooding, which is omitted from these studies, all of the selected cities are facing fluvial (i.e. riverine) flooding due of overtopping of embankments, pluvial flooding due to local rainfall events or both. Retrospective studies of past events cover for instance flood events in Dhaka (Gain et al, 2012),  Guangzhou-Shenzhen (Chan et al, 2015) or Calcutta, Delhi and Mumbai (De et al, 2013). Based on a quick-scan of literature, news-items and data, the cities have been subdivided into three classes:

- Fluvial and pluvial:  Calcutta, Mumbai, New Delhi, Lahore, Jakarta, Manila, Seoul, Dhaka, Ho Chi Minh,  Lagos;
- Fluvial: Cairo;
- Pluvial: Beijing, Guangzhou-Shenzhen (Chan et al, 2015), Shanghai, Karachi, Mexico City, Tehran (flash floods), Istanbul

Besides urban growth, climate change and a range of other factors that affect the future flood risk of these cities, four of the cities are experiencing significant subsidence: Ho Chi Minh City, Jakarta, Manila and Shanghai (Deltares, 2005). Also Calcutta and Dhaka are coping with subsidence, but at a more modest rate.

# 3. Memetic algorithm optimised urban growth model

This chapter is based on:

Veerbeek, W., Pathirana, A., Ashley, R., & Zevenbergen, C. (2015). Enhancing the calibration of an urban growth model using a memetic algorithm. *Computers, Environment and Urban Systems*, 50, 53-65.

## 3.1 Introduction

Originally used in regional economics (e.g., Almeida, 1954; Alonso, 1964; Forrester, 1969), land use change models, particularly urban growth models, attempt to mimic historical and future LULC transitions in a spatially explicit manner. Depending on the selected representation of the spatial components, cells or patches represent discrete LULC classes (i.e., states) that can change over time. These changes are affected by a set of drivers that are conceptualized as transition rules. Contemporary urban growth models are often based on cellular automata (CA) models (e.g., White et al, 1993; Batty and Xie, 1994; Clarke et al, 1997; Li and Yeh, 2000) that describe cell-based LULC transitions as a function of local interactions, which represent neighbouring conditions that drive the formation of spatial urban patterns. Often, these models are combined with 'top-down'-driven transition rules that incorporate fixed physical properties (e.g., slope or elevation) and/or statistically determined growth drivers (e.g., population growth or economic development). Although most models were essentially developed as generic models capable of representing the growth dynamics of any metropolitan area, they can be adjusted to mimic LULC transformations in specific cities or regions. The model calibration and validation stages can be performed manually, but they are frequently automated using historical LULC data as a training set (e.g., Li and Yeh, 2002). During calibration, the relation between the predicted and observed LULC, by using a set of metrics determining the goodness-of-fit (e.g., Næsset, 1995), are combined with an update function that changes the transition rules. When an optimal correlation is found (i.e., no significant improvement in the goodness-of-fit can be obtained), the growth transition rules are applied to prospective years to obtain the projections. This process is depicted in Figure 3.

Machine-learning algorithms or other regression methods are frequently used to cali-

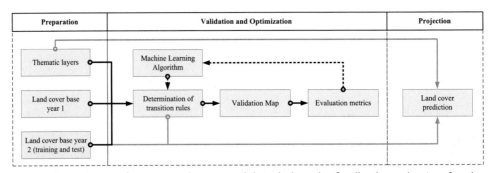

**Figure 3:** Typical setup for a LULC change model, including the feedback mechanism for the calibration

brate LULC change models. Long et al (2009), as well as many other authors (e.g., Liu et al, 2003; Hu and Lo, 2007), used logistic regression to optimize transition rules. Li and Yeh (2004) applied an artificial neural network to optimize parameters, while Yang et al (2008) used a support vector machine. Recent applications include particle swarm optimization methods (Feng et al, 2011; Rabbani et al, 2012) or ensemble learning strategies in which several machine learning algorithms are executed in parallel (Gong et al, 2012). All of these stochastic methods are commonly set up as iterative processes that require multiple model runs to obtain convergence to satisfactory solutions. Such approaches work reasonably well when the computational costs of running a calibration and validation sequence are relatively low. For example, when a LULC transition for single cells is calculated, the result is evaluated and an adjustment is made to one of the controlling parameters; thus, the computational requirements are minimal. Yet, some models rely on both local and global comparisons of the LULC change maps for evaluations. Particularly large areas composed of millions of cells are computationally very costly, resulting in a long-duration calibration. While for instance GPU-accelerated calibration (Blecic et al, 2014) could cope with the increasing demand for computational power, the underlying methods do not fundamentally change. In addition, while calibration using machine-learning algorithms can produce LULC changes that mimic observed transitions, overfitting the parameters might lead to variable future projections. Thus, the observed spatial development trends in historical data are discontinued when running many subsequent iterations of the calibrated LULC change model.

By building upon a Dinamica-EGO-based LULC change model (Filho et al, 2003; Filho et al, 2009), a 2-stage modelling approach is introduced to separate the calculation of the urban-area growth from the diversification of the growth extent into urban LULC classes. This method ensures the production of consistent LULC patterns over long periods. The model is equipped with a customized automatic calibration method based on a genetic algorithm (GA). The GA is extended with a local search function, which significantly reduces the required number of candidate solutions and iterations to produce robust and accurate results. This approach provides an alternative for the often used 'off-the-shelf' machine-learning algorithms used in LULC change models. To test the outcomes, the model is initially applied to the Beijing metropolitan area, which is an ideal case study due to the combination of market-driven rapid urban expansion and top-down planning policies (Han et al, 2009). The model is required to adjust to alternative urban development trends that might not simply evolve near the current urban clusters. Furthermore, the relatively large case study area, combined with the applied 30-m spatial resolution, could require substantial computations.

In the first section of this chapter, the case study and the Dinamica EGO model are introduced, including a detailed description of the methods used to define transition rules and the evaluation criteria used to estimate the goodness-of-fit of the produced LULC changes. The second part provides the background and context for the development of the 2-stage approach, as well as the GA extended calibration. Subsequently, the outcomes are presented. A comprehensive analysis should provide sufficient evidence for the robustness of the observations and interpretations. Finally, a brief discussion of the underlying assumptions and ongoing issues is presented. Most of this chapter has been previously published in Computer, Environment and Urban Systems (Veerbeek, 2015).

## 3.2 The Case Study

### 3.2.1 Beijing

The case study used to test the model is greater Beijing, China. To an extent, Beijing's urban development is typical for an Asian megacity; since the late 1980s, the city has undergone massive expansion and redevelopment which has doubled the size of the urban extent in the last 15 years. Over 1995-2005, this expansion comprised 19% infill, 75% extension and 6% leapfrogging development (see Figure 4). Although greater Beijing is surrounded by mountains to the north and west, the city's potential for development is relatively unconstrained. However, Beijing's future growth is not unlimited; the absence of freshwater bodies (Yong, 2009) and the increasing traffic congestion (Zhao, 2010) are likely to limit Beijing's expansion in the long term. Despite the large degree of freedom for development, Beijing remains relatively compact. The majority

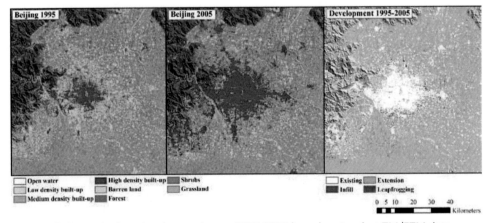

**Figure 4:** Estimated urban development over 1995-2005 based on Landsat TM/ETM data

of the urban extent is contiguous and expanded from the 15th century "Forbidden City", which forms the geographic centre of the city. Nevertheless, Beijing is facing significant expansion due to urban sprawl (Zhao, 2010), which has emerged over the last decade.

In 2005, the Beijing metropolitan area housed approximately 15 million inhabitants (Beijing Statistics Bureau, 2005), which had been expected to increase to 18 million by 2020. However, the current population has already surpassed 19.6 million (National Statistics Bureau, 2011), and the Beijing Academy of Social Sciences recently revised their estimate to 26 million by 2020 (Caixin Online, 2012). These discrepancies show that there is no real consensus regarding the population growth in Beijing and that future containment might be difficult to achieve through policy and urban growth constraints. In contrast to many other rapidly developing megacities, the urban development of Beijing is being facilitated through a succession of regional development plans and urban master plans (Beijing Municipal Planning Committee, 2006) that specify particular development regions, functional zones and excluded areas. Yet, Han et al (2009) showed that more than 35% of the urban development in Beijing had exceeded the planned urban extent by 2005. Long et al (2012) estimated that a 51.8% exceedance would occur between 1991 and 2010. Furthermore, long-term planning policies change over time, e.g., the former Beijing greenbelt policy initially proposed in 1993 and extended in the 2005 general plan had only limited success. Yang and Jinxing (2007) show that the first greenbelt failed to exclude the urban development of the city. Nevertheless, the relatively strong top-down planning regime of Beijing could provide a challenge to growth models that are exclusively based on past LULC data because planned future growth containment and development zones are not explicitly included in the model. Planned development areas that do not conform to previous spatial development trends might be missing from the present future projections. In contrast, given the previous discussion, the value and level of control exerted by the Beijing planning policy may be questioned; the current development pressure in the area might lead to urban expansion based on optimal suitability instead of top-down planning and to inconsistent, urban containment zones. Another challenge when developing an LULC change model for Beijing is the limited availability of data. Detailed spatial datasets are only available to municipal departments or Chinese institutes; they will not be made available to external researchers or users. Therefore, data can only be derived from limited, publically available sources. These data might be outdated and provide only a limited level of detail given the aim of this study.

Urban growth models for Beijing have been developed by He et al (2006; 2008) and Long et al (2009). Both models use extensive datasets, including top-down planning data from the existing master plans, planning constraints (e.g., protected areas) and other resources, that shape the Beijing urban planning policies. Unfortunately, the various models use different geographic extents, resolutions, horizons and validation metrics that complicate comparisons and cross references.

## 3.3 Data and model setup

For this study, the Dinamica EGO modelling platform (Filho et al, 2003; Filho et al, 2009) was used to simulate LULC transformations in the case study areas. This platform provides a flexible and well-tested environment for dynamic GIS modelling, including a set of geoprocessing tools specifically developed for modelling LULC changes. One of the main advantages of using Dinamica EGO is that it applies more conventional and established methods of statistical estimations to a CA-based approach in a multivariate framework (Liu, 2009: 50). This produces a model with a greater level of control than a traditional CA-based LULC change model in which spatial patterns are predominantly controlled by local interactions. These processes have important consequences for the calibration of the model, in which the optimization focuses on parameter sets that differs from traditional CA-based LULC change models. The performance of Dinamica EGO in the context of LULC change modelling has been evaluated extensively and most recently by Mas et al (2014).

Yi et al (2012) compared a Dinamica-EGO-based LULC change model for China with outcomes produced in a CLUE-S model. Their conclusion was that while the LULC patterns produced by both Dinamica EGO and CLUE-S broadly matched those of the observed changes, Dinamica EGO better predicts the actual amount of land use change. Although these outcomes highly depend on the applied input data and calibration procedure, they do provide some evidence for the validity of Dinamica-EGO-based LULC projections.

Data processing was performed on Landsat 5 TM and ETM images (USGS and NASA, 2009) to classify the land cover. To develop additional thematic layers, ASTER GDEM terrain data and recent Open StreetMap data for infrastructure were used. To obtain a coherent dataset, raster data maps were reprojected and resampled to a resolution of 30 m using the derived land cover maps as references. The base years were set as 1995 and 2005 to obtain representative growth rates for the past decade. A maximum-likelihood-based multi-temporal land cover classification (Bruzzone and Serpico, 1997) was applied by extending the classifications of the base years by two intermediate years (1997 and 2002). Additional error corrections were performed manually using Google Earth™ multi-temporal imagery. For the classification, the NLCD 2001 Land Cover Class

Definitions (Homer et al, 2004) were used, from which built-up areas were divided into 3 density levels: low, medium and high. Heights and slopes were calculated from the ASTER GDEM sets and were added as a separate layer. Because the road network was assumed to remain static during the calculation of the future LULC projections, only regional and truck roads were maintained to prevent possible bias in the growth patterns toward existing local road structures.

These datasets were divided into three parts: (i) LULC maps for the base years; (ii) static thematic feature maps; and (iii) normalized static proximity maps. The static thematic maps include elevation and slope data, while the normalized proximity maps contain the distances to the surface water/stream network and main roads. The road network was subdivided into sets of urban roads and regional roads. This division is based on the contour provided by the urban footprint, which is calculated by applying a method adopted from Angel et al (2007). All distances were conceptualized using a weighted cost function, where the slopes derived from the DTM act as weights. Apart from these maps, a set of dynamic variables were calculated to determine the distances of individual LULC cells to specific features (e.g., distance to forest).

Because only limited data were available from public sources for the Beijing study area, one of the main challenges in this project was to derive as much information as possible from the existing datasets. A major driver for urban growth has been the proximity to the central business district and to other more local urban clusters (e.g., He et al, 2008; Hu and Lo, 2007). To identify the regions adjacent to large urban clusters, the commonly used kernel density (Bailey and Gatrell, 1995) is calculated. By applying a relatively large radius, local features were obscured in favour of large contiguous built-up areas. The periphery of medium- and small-sized urban clusters was identified by using the patch-size distribution of all built-up areas. This method relies on the assumption that the majority of built-up patches consist of small clusters, and only a very small subset contains large contiguous built-up land. Thus, by ranking the patches according to size, a distance map to patches was produced, with the exception of the largest 5%: $r_i > 0.05N$, where N is the total number of built-up patches and $r_i$ is the rank number of patch $i$ in an ordered set of decreasing patch sizes. Both the kernel density and the patch size distribution were used as additional dynamic maps for estimating the LULC transitions.

LULC change models in Dinamica Ego often rely on the 'weight of evidence' method to estimate the relative importance of thematic features in the observed LULC transitions. The implementation is adapted from Agterberg and Bonham-Carter (1990) and Goodacre et al (1993). Originally, this method was developed to calculate empirical relationships of spatial variables as represented by categorical or continuous map data.

To develop these weights, all map values (e.g., slopes) and variables (e.g., distances) are first subdivided into sets of ranges. Generally, the intervals are based on calculating the geometric intervals, natural breaks or quantiles in the observed distributions. These provide binary patterns that act as variables from which the effect on LULC transitions is calculated by using a Bayesian probability function. A single LULC transition described as event D, given a binary map B that defines the presence or absence of a pattern (i.e., the interval between two values), is expressed as a conditional probability:

$$(1) \qquad P\{D|B\} = \frac{P\{D \cap B\}}{P\{B\}}$$

where $P\{D \cap B\}$ describes the number of occurrences of $D$ given $B$ divided by the number of cells $D$ in a raster map. $P\{B\}$ represents the fraction of the area occupied by pattern $B$ divided by the area occupied by the map extent. Transformation of equation (1) to express the prior probability of $D$ and rewriting it into a logit form obtains the positive weight of evidence $W^+$ of occurrence of $D$:

$$(2) \qquad \ln\{D|B\} - \ln\{D\} - W^+$$

$W^-$ describes the negative weight of the evidence:

$$(3) \qquad W^- = \ln\left(\frac{P\{\bar{B}|D\}}{P\{\bar{B}|\bar{D}\}}\right)$$

where $\bar{B}$ is the absence of the occurrence of pattern $B$ and $\bar{D}$ is the absence of the land cover transition. If the occurrences of $D$ were observed more often than would be expected due to chance, then $W^+$ would be positive and $W^-$ would be negative. This method is extended to accommodate multiple spatial patterns, i.e. $B, C, D, \dots N$

$$(4) \qquad P\{D|B \cap C \cap D \dots \cap N\} = \ln D + W_B^+ + W_C^+ + W_D^+ + \dots + W_N^+$$

Because the sum of the prior odds ratio of and $\bar{D}$ equals 1, the odds ratio can be replaced by:

$$(5) \qquad P\frac{\{D|B\}}{1 - P\{D|B\}}$$

This equation leads to the post-probability calculation of a transition of land cover class $i$ to $j$, given the spatial pattern $B, C, D, \dots N$ at location $(x,y)$:

$$(6) \qquad P\{i \Rightarrow j \mid B \cap C \cap D ... \cap N\} = \frac{e \sum W^{+}}{1 + e \sum W^{+}}$$

An example of an obtained weight distribution is illustrated in Figure 5, showing the influence of the distance to the main road network for the LULC transition from grassland to low density built-up areas. By estimating how strong this influence is within all observed LULC transitions between the base maps, the weights were established. This estimation process was conducted for all ranges within the thematic map layers and variables to provide a transition probability map for all observed LULC transitions. The derived weights can also have negative values. Therefore, apart from driving forces, resistance forces to urban expansion could also be expressed.

One of the main advantages of the weights of evidence method is that, given that the spatial variables are independent, the effect of each spatial pattern on a transition can be calculated independently of a combined solution. The number of calculations required to determine the weights is therefore not combinatorial, e.g., a set of 7 land cover transitions using 5 spatial variables divided into 10 ranges of values require a calculation of 7 x 5 x 10 = 350 weights. To ensure independence of the spatial variables, Dinamica EGO offers a pairwise test for calculating the chi-square-based Cramers and contingency coefficient (e.g., Almeida et al, 2003) and the joint information uncertainty based on the joint entropy measure. This latter metric is extensively described in Bonham-Carter (1994).

The effectuation of the LULC changes was conducted using 2 stochastic CA modules: the Expander and Patcher. The Expander aims to expand or contract existing patches of cells, thus mimicking urban extension, while the Patcher initiates isolated new patches (i.e., leapfrog development). The characteristics of the CA in the Expander and Patcher

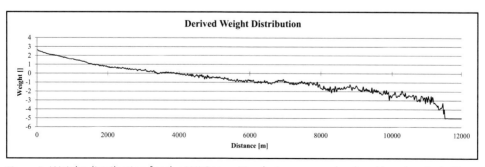

**Figure 5:** Weight distribution for the LULC transition between grassland and low density built-up areas as a function of the distance to the main road network

modules are extensively covered by Almeida et al (2003). The technical setup of both modules is similar, with the exception of the transition locations within the CA neighbourhood. In the Expander module, candidate transitions can only be located adjacent to a cell of the target LULC class, e.g., the transition of barren land to high-density built-up area can only occur next to a high-density built-up area. In the Patcher module, however, the base cell cannot be located adjacent to a cell with the target LULC class. Both CA models work probabilistically and use a pre-determined neighbourhood in combination with the calculated probability map based on the weight-of-evidence. Cell transitions were distributed based on the probability rank within the neighbourhood. A so called 'prune factor' determines to what extent the transitions are ranked by introducing a stochastic factor. A higher level of stochasticity means that transitions with a lower probability are included as candidates for the LULC transitions. The rate of transition is determined by calculating the net transitions between LULC classes between the two base years. These transitions are distributed over the two CA modules using a third module: a Modulator, in which the distribution rates between the Expander and Patcher modules are set for every observed LULC transition; effectively, the CA modules act as a constrained CA (Engelen et al, 1997).

Model validation was performed using a neighbourhood-based Kappa index (e.g., Hagen, 2003) with a constant function that measures the goodness-of-fit between the predicted and observed LULC change base-year interval; a fuzzy comparison function was used for different window sizes. The minimum mean similarity (Soares-Filho et al, 2009) was based on the fuzzy similarity (Hagen, 2003), and the concept of fuzziness of location was used to establish whether two cells of a map category (i.e., land cover class) overlapped. Fuzziness of location is required overcome the problem of evaluating shifted spatial patterns as different (e.g., when comparing 2 checkerboards patterns, where the black and white fields are inverted on one board). The similarity measure for a pair of maps is obtained by performing a cell-by-cell fuzzy set intersection:

(7) $$S(V_A, V_B) = \left[ \left| \mu_{A,1}, \mu_{B,1} \right|_{Min}, \left| \mu_{A,2}, \mu_{B,2} \right|_{Min}, \ldots, \left| \mu_{A,i}, \mu_{B,i} \right|_{Min} \right]_{Max}$$

where $V_A$ and $V_B$ are are the fuzzy neighborhood vectors for maps $A$ and $B$, and $\mu_{A,i}$ and $\mu_{B,i}$ are the neighborhood memberships for land cover classes in maps $A$ and $B$. To overcome overestimation from the one-way comparison (e.g., the spatial pattern in one map is a subset of the pattern in the other map and therefore shows a 100% correspondence), a two-way comparison was introduced from which the minimum mean similarity was derived, i.e., the 'minimum mean similarity'. This process is required because only the land cover changes were compared, rather than the resulting land cover map. The window size used in the fuzzy comparison was variable (Constanza,

1989) was thus able to provide a goodness-of-fit at different resolutions. MMS values are expressed within the interval of $[0, 1]$, where 0 indicates a 0% correspondence and 1 indicates a 100% correspondence between the predicted and observed LULC maps. The outcomes presented in this paper mostly relied on a window size of 5, which corresponds to a resolution of 150 (5 x 30 m).

## 3.3.1 Model refinement: Sequential approach

While a satisfactory goodness-of-fit might be achieved during model validation, consistency of the spatial characteristics of the produced spatial patterns in the projection stage is not a guaranteed. When running multiple iterations, spatial trends might appear that were not initially noticeable. These trends might be caused by overfitting the model to the base maps for the calibration. Subsequently, the parameterization of the model failed to describe the actual mechanisms that produced the LULC transitions; instead, it produced a validation map based on random error.

A potential cause of the inconsistent projections could occur due to the division of the built-up areas into 3 density levels, which significantly adds to the model complexity. While improving the expressiveness of the outcomes, this division introduces additional LULC transitions between both rural and built-up LULC classes, as well as within the built-up LULC classes. Instead of a single probability for a transition of a given rural LULC class to a single LULC class for built-up areas, a set of three transitions is available that might score nearly even probabilities. Because of the winner-takes-all principle, the highest ranked transition is executed, even though the calculated probability of other rural-urban LULC transitions might only be marginally lower. The likelihood of incorrect LULC transitions therefore increased. Arguably, the outcome could be regarded as similar: low-, medium- or high-density built-up areas can all be aggregated into a single built-up class. Yet, the subdivision might create an undesirable side-effect, i.e., propagation of errors in the spatial distribution (i.e., the locations) of the rural-urban LULC transitions due the conditional independence of the transitions and the associated weights-of-evidence. No specific characteristics in the transition or the topology, e.g., high-density urban patches, are provided. A typical densification pattern, as often observed along the fringes of cities, is therefore not specifically provided; rural-urban LULC transitions are therefore ill-defined, and a process occurs where distributed small pockets of built-up areas attract rural-urban LULC transitions at a nearly equal rate to large contiguous urban areas. Although this process is somewhat dampened by the application of the earlier described kernel density and patch-size driven dynamic maps, the subsequent weights were not large enough to facilitate a more compact formation of built-up areas. In practice, this process could lead to substantial dispersion and fragmentation of built-up areas in the projection stage, i.e., after running enough iterations for the errors to become significant.

The level of fragmentation in built-up areas can be assessed by calculating the fractal dimension (FD), which is expressed by a value in the interval [1,2] to assess the complexity of the observed urban patterns (e.g., Turner, 1990; Herold et al, 2005). Values approaching 1 signify shapes with relatively simple perimeters (e.g., squares), while values approaching 2 signify highly intricate shapes. If the FD for the projections fits within the extrapolated range calculated for the validation base-year interval, then the projected spatial patterns confirm the observed historical spatial trends. Otherwise, the produced distributions are inconsistent by showing either a disproportionate amount of urban sprawl or clustering signified by a substantially higher or lower FD. In the current base years used for model calibration, the FD drops from 1.52 in 1995 to 1.51 and 1.48 in 2005 and 2010, respectively; these values indicate a tendency toward a marginally more compact distribution of built-up areas in the Beijing metropolitan area.

To overcome this problem, a sequential approach was developed to produce the LULC projections:

1. *'Urban envelope'-development.* In this stage, the future urban extent is projected using a single urban LULC class, i.e., all urban LULC classes are aggregated into a single class;

2. *Urban diversification.* The urban envelopes are used as spatial development constraints for the development of multiple urban LULC classes. Transitions to different urban LULC classes only occur in cells that are members of the envelope defined in the first stage.

The first stage forces the model to behave as a spatially constrained LULC change model. Yet, contrary to most constrained models (e.g., Long et al, 2009) that integrate top-down planning policies to define the actual growth boundaries for future projections, this approach merely defines containment boundaries by running a simplified version of the LULC change model, in which the issue of the conditional independence of rural to urban LULC transitions is avoided. The advantages of this approach are that the urban development becomes more bounded and the simulated LULC changes associated with urban expansion and/or densification are better defined. The actual subdivision into a given set of subclasses expressing different densities of built-up areas is then limited to projected 'urban envelopes', which therefore limit the potential errors of misclassification within the range of the built-up area LULC classes (e.g., medium instead of high-density built-up area).

## 3.3.2 Automated Calibration

One of the main factors for obtaining a high goodness-of-fit for the model setup de-scribed above is the parameterization of the Modulator, Expander and Patcher mod-ules that were used to determine the growth of the existing urban areas and the for-mation of new urban areas. All 3 modules comprise extensive parameter sets that determine the rate, size, variation, isometry and potential candidates for LULC transi-tions. Although manual calibration of the parameters could lead to good results, ex-tensive experience and trial runs are required to achieve an acceptable goodness-of-fit between observed and predicted LULC distributions during validation. Because the pa-rameters are conceptualized as continuous variables, a random or brute-force search might result in suboptimal goodness-of-fit or extensive computational costs.

An increasing number of researchers have applied genetic algorithms (GA) for auto-mated calibration, possibly due to the relative simplicity of the methods (e.g., Li et al, 2007; Shan et al, 2008). Typically, a GA consists of a population of randomly initialized candidate solutions (i.e., parameter sets with different values) that, through a process of cross-over, mutation and selection, improve their performance over a range of gen-erations. For an LULC change model, this would typically mean that every candidate solution represents (part of) the transition rules. The goodness-of-fit that is used for validation ranks the performance of the candidate solutions, after which a selection mechanism is applied to select the best-performing individuals. Through cross-over and mutation, the parameter sets are then adjusted to serve as a new generation; then, the process is repeated.

Although GAs are successfully used as optimization algorithms for a large range of problems (e.g., Eiben and Smith, 2003), they generally require a large population of candidate solutions and a large number of iterations to reach a global optimum with-in a search space. Because the evaluation function (i.e., the validation sequence) for the application of the Dinamica-EGO-based LULC change model is computationally ex-pensive, using a population set of 100s or even 1000s of differently parameterized candidate LULC change models is not feasible, particularly when applied to the study area in this project in which more than 25 million cells are included. The cross-over

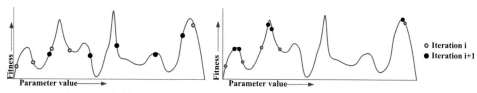

**Figure 6:** Exploration (left) and exploitation (right) in a GA

and mutation rates in a GA largely determine how well the algorithm explores a given search space and how well the solution converges to an optimum; this is typically referred to as the problem of exploitation and exploration (Eiben and Smith, 2003), which is illustrated in Figure 6. In a one-dimensional search space containing a single global optimum, the illustration on the left shows a larger coverage of the parameter space (i.e., exploration), while the illustration on the right depicts optimization around local optima (i.e., exploitation).

The mutation rate often handles the exploration of the search space by introducing a stochastic factor in the parameter values of the candidate solutions, while the crossover mechanism handles the convergence by combining or averaging parameter values of the candidate solutions that show a good performance. Generally, GAs excel at rapidly identifying good solutions (i.e., exploration) but are often less suited for finding an optimal solution (i.e., exploitation) (Eiben & Smith, 2003). To overcome this problem, a so-called memetic algorithm (MA) (e.g., Ong et al, 2004; Chen et al, 2010) has been developed, in which the standard GA is extended with a local search algorithm (i.e., a 'hill climber'). The concept is that the GA ensures the exploration and coarse exploitation of the calibration, while the local search fine tunes the candidate solutions to reach local optima within the search space. Although this approach does not guarantee that the optimal solution will be found, convergence to higher fitness values might proceed more effectively. The addition of a local search algorithm to enhance the performance of a machine-learning algorithm is not necessarily limited to the domain of GAs but extends to other branches, including support vector machines (e.g., Nekkaa and Boughaci, 2012) and particle-swarm optimization algorithms (e.g., Mousa et al, 2012). Consequently, MAs can be regarded as extensions of 'regular' machine-learning algorithms, although this distinction is fading with the introduction of nested and ensemble models (Zhang and Ma, 2012).

As already noted, within a Dinamica EGO LULC change modelling context, the optimization addresses the set of meta parameters that complement the weight of evidence based on transition rules. The parameters do not focus on the individual cell level of the model but address regional and global characteristics of the produced patches. The evaluation of such characteristics can only be performed using the complete model outcome, as opposed to evaluating the observed and predicted transitions of individual cells, as seen in the calibration of many other LULC change models, e.g., IDRISI (Clark Labs, 2012).

In the current implementation, the GA and the MA are applied to optimize the parameter values controlling the Modulator, Patcher and Expander modules that facilitate the LULC transitions. These modules consist of three parameter sets that control processes, or independent optimization problems, in sequential steps:

1. *Modulator parameters.* This step determines the optimal distribution of transitions between the Expander and Patcher modules; in practice, this equates to the rate of extension and leapfrogging of built-up areas;

2. *Patch characteristics.* Optimization of the patch size, variance and isometry determines the quantitative characteristics of cluster (i.e., patch) LULC transitions;

3. *Neighbourhood characteristics.* The neighbourhood determines the area where the land cover transitions are considered, i.e., the area for which the ranked probability is determined for land cover transition A to B. Additionally, the prune value is optimized by determining the level of stochasticity in prioritizing the rank of the candidate transitions.

The implementation of the local search function is straightforward: a scalar $A$ operating on a vector $V = \langle x_1, x_2, \ldots, x_n \rangle$ represents a set of parameter values.

The applied GA uses 2-point arithmetic crossover, which was conceptualized as follows: Let $V_a = \langle x_1, x_2, \ldots, x_n \rangle$ and $V_b = \langle y_1, y_2, \ldots, y_n \rangle$ represent parameter vectors (e.g., the modulator ratios). Let k and l be two random positions where $l > k$ and $l \leq n$ Between positions $k$ and $l$, the values in the offspring $C_a$ and $C_b$ are are calculated as the arithmetic means of the entries in the two parents:

$$(8) \quad C_a = \left\langle x_1, \ldots, x_{k-1}, \frac{x_k + y_k}{2}, \ldots, \frac{x_l + y_l}{2}, x_{l+1}, \ldots, x_n \right\rangle$$

$C_b$ is created in the same way but x and y are reversed.

Additionally, 2-point uniform mutation is implemented by replacing the entries between positions k and l by a set of random values. The range of these values depends on the actual parameter they represent. For instance, the ratio for the modulator change module should be in the range of $[0,1]$, while the isometry parameter should be chosen in the range of $[0,2]$. Other parameters have no upper bounds but are all larger than 0 because they represent area or neighbourhood sizes.

Parent selection is fitted proportionally using:

$$(9) \quad p_i = \frac{f_i}{\sum_{j=1}^{N} f_j}$$

where $f_i$ is the fitness of individual $i$ and $N$ is the total number of individuals, i.e., candidate solutions. To ensure that the best-performing candidate solution is always chosen for the generation of offspring, elitism is introduced: $\max_{i \in N} P_{rank}(i) = 1$, where

**Table 1:** Steps and parameters used for calibration

| Module | Variables | | |
|---|---|---|---|
| | Stage 1 | Stage 2 | Stage 3 |
| Modulator | Modulation Rates [0,1] | | |
| Patcher/Expander | | Mean Patch Size[ ha] | Neighbourhood Size [] |
| | | Patch Size Variance [ha] | Prune Factor [] |
| | | Patch Isometry [0,2] | |

$\max P_{rank}$ represents the best-performing candidate solution.

For every LULC transition, a set of 11 parameters control the formation of patches. Because 7 types of LULC transitions were identified for the Beijing study area, the total number of parameters to be optimized was 77. However, the observed transition rates were not equally distributed; the majority of LULC transitions consisted of barren land to built-up areas (75%) and grassland to built-up areas (17%). Thus, optimizing the parameters for these transitions should have the largest effect on the goodness-of-fit between the predicted and observed LULC changes. An overview of the steps and parameters is provided in Table 1.

As seen in Table 1, the parameters often had different ranges. Therefore, the increments for the local search differ per parameter.

The implementation of the local search algorithm as part of the calibration and validation sequence is depicted in Figure 7. Note that the introduction of a local search algorithm introduced a new loop within the calibration sequence. This sequence leads to additional runs of the LULC transition model, which can be computationally expensive. A summary of the previously described characteristics and methods applied in the GA and MA are presented in Table 2.

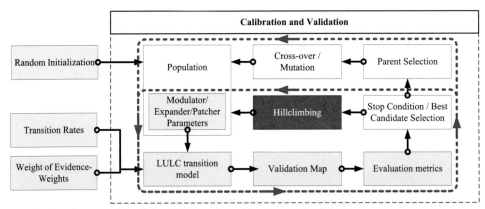

**Figure 7:** Implementation of the MA into the calibration and validation sequence.

**Table 2:** Summary of the main characteristics of the applied GA and MA schemes

| | |
|---|---|
| Representation | Floating point |
| Recombination | 2-point arithmetic |
| Mutation | 2-point uniform |
| Parent Selection | Fitness proportional with elitism |
| Replacement | Generation based |
| Population size | 8; 32 |
| Search enhancement | Local search |
| Increment size | linear |

# 3.4 Outcomes

To estimate the performance of the MA optimized calibration, a baseline was created by applying a 'standard' GA using the previously described setup with a population size of 32. The stop criterion was set at 96 iterations, which proved to be sufficient because no significant performance gain (i.e., MMS increase) could be observed. To avoid over-calibration or overfitting, the LULC base maps for 2000 and 2005 were used for calibration (i.e., training) and an additional LULC map for 2010 was used for validation (i.e., testing). This setup was repeated using the 3-step MA optimized calibration method, where the number of iterations for the first two steps was set as 8 per step. The stop criterion was set at the third step at 16 iterations because no significant performance increase could be observed. Because one of the assumptions is that the local search algorithm in the MA will reach local optima during every iteration, an additional configuration was tested with a limited population size of 8 candidate solutions. To ensure statistically significant outcomes, both procedures were repeated 32 times. The outcomes are presented in Figure 8, which illustrates the progression of the mean and 5th-95th percentile ranges for the observed maximum MMS values over the iterations. The final values are summarized in Table 3. To ensure comparability, the same random seed was used within the CA modules.

The comparison of the outcomes focuses on the fitness development over the iterations and the end results:

- Fitness development. The graph showing the overall GA-produced progression of MMS values is typical for GA-based optimization processes: a curve-like performance increases over the range of iterations starting with a relatively large initial fitness increase that flattens after approximately 64 iterations. The shape of the MA-optimized calibration looks somewhat different. The subdivision of the MA-optimized calibration into 3 separate steps is particularly influential in the first two steps in which, after an initial moderate increase in

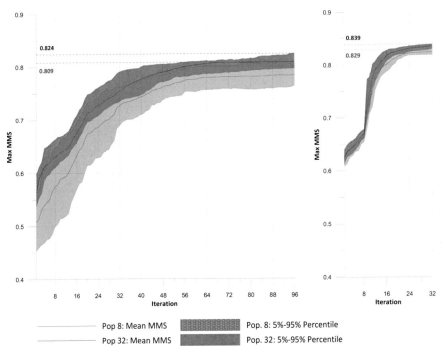

**Figure 8:** Mean and 5th-95th percentiles for the observed maximum MMS values using GA-enhanced (left) and MA-enhanced (right) calibrations.

the MMS, a considerable increase is observed after the 8th iteration.

- Minimum, Mean and Maximum MMS. As shown in Table 3, application of the MA shows a marginal increase in the observed maximum MMS when compared with the GA-optimized calibration: 0.015 and 0.024 for population sizes of 32 and 8, respectively. For the observed minimum and mean MMS, the differences are somewhat more substantial, particularly when comparing the results for a population size of 8.

- Required number of iterations. The MA shows a significant performance gain over the GA; after 32 iterations, no significant performance improvement is observed for the MA, compared with 96 for the GA-enhanced calibration. Furthermore, the maximum MMS achieved for the GA is already attained after 19 and 18 iterations by the MA for population sizes of 32 and 8, respectively.

- Variability. While the variability in the observed MMS values for the GA increases significantly for a smaller population, the variability within the distribution produced by the MA-optimized calibration remains nearly equal.

The observation concerning the variability is particularly important because it suggests that a consistently high performance can be achieved with only a small population

**Table 3:** Mean, 5th, 95th percentiles and resulting range of MMS values after 96 and 32 iterations for the GA and MA optimized calibrations, respectively.

|  |  | 5th Perc. MMS | Mean. MMS | 95th Perc. MMS | Range |
|---|---|---|---|---|---|
| GA | Pop. 32 | 0.796 | 0.809 | 0.824 | 0.028 |
|  | Pop. 8 | 0.763 | 0.784 | 0.809 | 0.047 |
| MA | Pop. 32 | 0.831 | 0.835 | 0.839 | 0.008 |
|  | Pop. 8 | 0.819 | 0.825 | 0.829 | 0.009 |

and after a significantly low number of iterations. To determine if this observation is obtained at different window sizes, i.e., at different resolutions, the MMS distribution is calculated when combining the best-performing solutions of the 32 individual runs. The outcomes are shown in Figure 9, where the MMS for a population size of 8 is shown for window sizes between 30 and 510 meters.

Based on Figure 9, particularly for larger window sizes, the variability in the MMS values is significantly higher when using the GA-optimized calibration. Additionally, at higher resolutions (particularly for smaller population sizes), the MA-optimized calibration has a higher probability of producing a large MMS.

One of the drawbacks of the MMS as a metric for estimating the goodness-of-fit be-

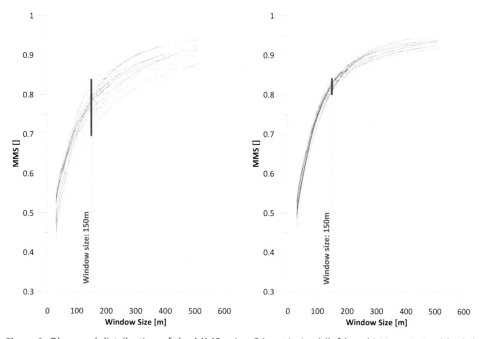

**Figure 9:** Observed distribution of the MMS using GA-optimized (left) and MA-optimized (right) calibrations.

**Table 4:** Confusion matrix and kappa index for the 1995-2010 transitions. The projected LULC cell changes are in the rows, and the observed cell changes are in the columns.

| | water | Low D built-up | Med. D built-up | High D built-up | barren land | forest | shrubs | grass-land | accuracy |
|---|---|---|---|---|---|---|---|---|---|
| water | 282020 | | | | | | | | NA |
| LD built-up | | 248883 | 271586 | 212049 | 131758 | 36 | 28489 | 200550 | 99.81% |
| MD built-up | | 226504 | 173653 | 351286 | 567337 | 785 | 16100 | 51249 | 98.74% |
| HD built-up | | 248510 | 330647 | 335669 | 161683 | 2289 | 1503 | 10732 | 99.14% |
| barren land | | 261601 | 444157 | 155019 | 2931529 | 1 | 12 | 1092 | 99.97% |
| forest | | 1042 | 1233 | 835 | 0 | 819250 | 0 | 0 | 100.00% |
| shrubs | | 10994 | 17298 | 17800 | 0 | 0 | 322402 | 0 | 100.00% |
| grassland | | 89230 | 145440 | 27860 | 0 | 0 | 0 | 1225436 | 99.93% |

**Kappa (k): 95.12%**

tween predicted and observed LULC maps is that it evaluates both the rate and location of LULC transitions, i.e., the absence of a transition and a transition at an incorrect location are equally penalizing. To distinguish between errors caused by incorrect transition rates and locational errors, a confusion matrix was produced that quantitatively compares the predicted and observed LULC changes. The outcome, including the commonly used Kappa index, is presented in Table 4.

The confusion matrix and Kappa index show that the net transitions between LULCs nearly equal the observed transitions. Yi et al (2012) also that concluded that the Markov process used in Dinamica EGO to estimate the transition rates for an arbitrary interval is capable of precisely predicting the amount of LULC changes. Therefore, the calculated MMS reflects the locational errors in the LULC transitions instead of the net transition.

The computational costs for running the MA-optimized model (population size 8, 32 iterations) on a I7, 4 core processor workstation with 16Gb of memory are approximately 46.4 hours, whereas the GA-optimized model (population size 8, 96 iterations) requires 55.68 hours. However, because robust outcomes for the GA-optimized model are achieved with a significantly higher portability when using a larger population (i.e., 32), the actual computational costs that guarantee comparable results are approximately 222.7 hours. This outcome indicates a reduction in the computational costs of approximately 80%. These figures largely depend on the map sizes used for the calibration, as well as on additional parameter settings, in determining the weight values. Therefore, the values cannot be regarded as fixed benchmarks. Yet, the observed reduction in the computational costs increases the usability of the calibration method for extensive case study areas at a high resolution.

The causes for the robust performance using the MA in a setting with a limited population size and number of iterations are evident. When applying the standard GA, exploration is limited because mutation only has a limited opportunity to develop alternative parameter values. Furthermore, because the candidate solutions are not explored during an iteration, exploitation (i.e., local optimization) is poor. The local search algorithm used in the MA seems to successfully drive the parameter values to local optima, resulting in a significantly better MMS. Importantly, there are exceptions to the earlier described observations. Although rare, the stochastic nature of the mutation and crossover parameters can in some cases cause the GA to outperform the MA. However, when the same random seed is applied for every stochastic parameter in the algorithms, the MA will at least provide an equal MMS, although the processing time is longer due to the additional processing time required for the local search.

## 3.4.1 No Free-Lunch

Although the initial results of applying a MA to the calibration of LULC change models are encouraging, the extension also adds complexity to the model by introducing a set of new parameters. These parameters consist of the step-size and the direction of the local search. If the step size chosen is too coarse, then the parameter value might overshoot the local optimum; in this case, the MMS could be lower than the initial MMS. The proposed candidate solution would therefore be discarded (i.e., no improvement in the goodness-of-fit is found). A step size that is too small would increase the MMS, but only at a significant computational cost, and thus reduce the effectiveness of the MA compared with a standard GA. When the direction of the local search (i.e., the sign of the scalar) is incorrect, the local search would deviate from the local optimum. Then, the solution is discarded again. Both of these issues are illustrated in Figure 10.

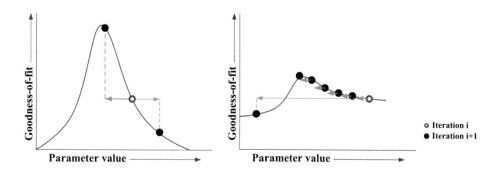

**Figure 10:** Problem of direction (left) and step-size (right) in local searches

In the implementation of the algorithm presented here, the issue of finding the direction was solved by repeating the local search in the opposite direction (i.e., the scalar has the opposite sign). In this case, an additional model run is required, which increases the computational costs. Although various solutions are available to overcome the problem of the step size and direction (e.g., Avriel 2003), the computational costs incurred by such approaches might be too high to justify the performance gain. Further research is required to test the performance gain by applying gradient descent (Snyman 2005) or heuristic functions (Battiti et al, 2008).

## 3.5 Projections

Using the MA-optimized model parameterization described previously, 12 LULC projections were produced at 5-year intervals through 2060. The outcomes are illustrated in Figure 11, including a focus on the built-up areas that were produced.

The most prominent observation from Figure 11 is the increasing dispersion of urbanized areas at global and local levels. In the geographically unconstrained areas east of Beijing, the projected urban patterns show significant fragmentation. Existing micro settlements (e.g., farms) seem to produce substantial extension and leapfrogging. This fragmentation can also be observed on a detailed level, where built-up areas are not adjacent to existing urban clusters but are sprawled around the fringes. Because these spatial trends require a significant number of iterations to propagate and become significant, the effect is limited in the validation year of 2010. Subsequently, the MMS expressing the goodness-of-fit is not necessarily compromised. However, the observed fragmentation seems counterintuitive because the urbanization observed in the base-year interval is mostly composed of compact expansion around Beijing's central urban areas. The corresponding FD for 2060 that would fit with this trend is close to 1.46 instead of 1.59 when calculated from the aggregated distribution of the built-up areas in Figure 11.

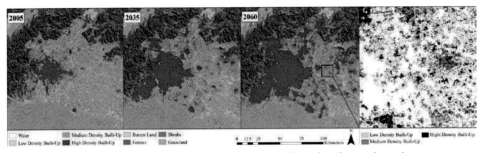

**Figure 11:** Projected LULC distributions for Beijing, including details on the urban areas produced for 2060.

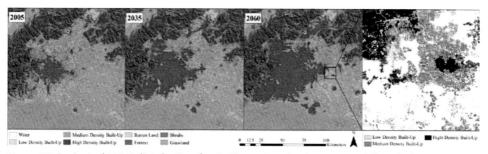

**Figure 12:** Projected LULC distribution for Beijing using the 2-stage model, including details on the urban areas produced for 2060 (bottom).2060.

Projections based on the previously introduced sequential approach, in which the projection of low-, medium- and high-density built-up LULC classes is limited to an 'urban envelope' projected in a prior stage, are presented in Figure 12.

The increasing levels of dispersion and fragmentation observed in Figure 11 are absent in the projections produced by the 2-stage model in Figure 12. The urban patterns are relatively compact and consistent with the patterns observed in the base-year interval. This finding is confirmed by calculating the FD, which is approximately 1.48 and 1.49 for 2035 and 2060, respectively.

Thus, without compromising the expressiveness of the model by limiting the LULC classes for built-up areas or compromising the goodness-of-fit during validation, the produced patterns seem to be better extrapolations of the observed spatial trends in the base years. This approach might be particularly useful for complex LULC change models involving many related LULC class transitions, where a single meta-class (e.g., built-up area) is subdivided into multiple sub classes. An additional advantage is that the approach also facilitates easy integration of spatially constrained top-down planning scenarios in which predetermined areas are assigned for urban expansion. These areas can be assigned to the base map containing the urban envelopes produced in the first stage. The actual urban development and diversification is then processed in the second stage without any modification of the model.

## 3.6 Discussion

Although LULC change modelling using Dinamica EGO might not necessarily pose the same challenges as other models, enhancing the optimization algorithms by using heuristics (i.e., experience-based 'shortcuts' that limit the search space for the algorithm to be covered) and hybrid approaches could provide further development of the often cumbersome task of model calibration. One of the main outcomes of the present work is that the adaptation of standard machine-learning algorithms exploit the characteristics of the calibration process of LULC change models and may significantly improve the model performance. Notably, the choice and implementation of machine-learning algorithms for calibrating LULC models often seem arbitrary and are too often based on standard 'off-the-shelf' tools. The conceptualization of the MA in the setup is not necessarily optimal. In the current application, the local search is implemented with a single scalar that works with a complete vector of parameters, albeit in three steps and for separate parameter sets. This generalized approach could be improved; however, there is a question as to what additional computational cost this would incur. This argument also extends to the application of an MA-optimized calibration to other LULC models. The approach requires extensive adaptation of the methodology to cope with the specific model characteristics and parameterization.

The utilization of various spatial characteristics (e.g., the kernel density) to extract additional information from the base maps and improve accuracy is at an early development stage. A vast array of statistical and geographic data can be extracted to extend the regularly used sources (e.g., distances to built-up areas and infrastructure or slopes). This notion extends to the applied evaluation method that in the current implementation is based on a single pattern-based metric. Incorporating for instance the FD into the evaluation function, could specifically target differences in the dispersion of built-up patches. Such an approach is taken by Li et al (2012), who combine three different spatial metrics to determine the goodness-of-fit between simulated and observed LULC maps. The outcomes show that by better capturing the spatial characteristics of the LULC changes, a significant reduction in the mean simulation error is achieved. Integration of multiple evaluation metrics, in future versions of the MA-enhanced model is relatively straightforward. Yet, additional research might be required to decide which metrics are most effective as well as understanding the relative importance of each of those metrics in the overall evaluation.

Splitting the production of LULC changes into two distinctive stages seems to increase the manageability and consistency of the projections in cases where LULC meta-classes consist of multiple subclasses. Yet, an alternative solution might be to develop additional dependencies between LULC transitions to better conceptualize spatial relations between, for instance, the transition to different types of built-up areas. Depending on the required expressiveness of future LULC models, further research on this aspect is required to cope with increasingly complex models.

A convincing claim against the validity of a projection exceeding 50 years is based on the short calibration period of a decade. The horizon mainly served to identify a technical issue: propose a possible solution to overcome inconsistencies observed for long-term projections using many subsequent iterations. To illustrate this issue clearly, the outcomes produced by a significant amount of model runs need to be included. Generally, the argument can be made that LULC change modelling currently lacks standardization and systemization; thus, many studies are ad hoc, and outcomes are often only useful within the scope of the particular research project. In this chapter, the presented material is not different; many of the claims need further testing and application in different case study areas. Fortunately, this work is foreseen in the scope of the CORFU project, which will prove if the presented approach leads to similar results in other megacities in South and Southeast Asia. Note that in this context, only data sources with global coverage are used as input data for the model.

In terms of usability, the model needs to be further developed. The model's current implementation as a Python script in combination with the Dinamica EGO platform is far from ideal. Interfaces with popular GIS platforms (e.g., Esri ArcGIS), as well as a graphic user interface, still need to be developed to make the application operational for a wide variety of third-party users. Furthermore, the application should be tested on a variety of cities to ensure validity, performance and applicability. Yet, all of these objectives have been well defined within the current scope of the project in which the model is being developed. A more thorough, step-by-step introduction of the model, as well as an application to additional case studies, a sensitivity test and a comprehensive analysis of the outcomes are provided in a technical report by Veerbeek et al (2014).

## 3.7 Conclusion

Automated calibration methods based on machine learning are successfully used for LULC change model parameterizations to produce high-quality estimations of observed LULC changes. However, overfitting of the models can lead to inconsistent long-term projections that show discontinuous spatial trends (e.g., increased fragmentation of built-up areas). Furthermore, when the computational costs of calibration are high, standard 'off-the-shelf' implementations of particularly stochastically based machine-learning algorithms could lead to extensive calibration times.

In this study, an alternative calibration method is presented for a Dinamica-EGO-based LULC change model using the metropolitan region of Beijing, China, as a case study. Instead of using a 'standard' genetic algorithm to optimize the vast parameter set, the algorithm was adapted to fit within a Dinamica-EGO-based LULC change modelling context and was extended by a local search function to facilitate faster convergence towards a locally optimal goodness-of-fit. The MA-enhanced calibration succeeded in providing robust, high-accuracy results with a significant reduction in computational costs. When compared with the observed land use and land cover changes, the calibration method achieved a pattern-matching accuracy of over 80% at a 150 m resolution. This accuracy was achieved using a small population size with only limited variability in the achieved accuracy. To cope with inconsistent projections over many iterations, the LULC modelling was divided into two stages. Initially, the model projects the 'urban envelope', a simplified LULC change model in which all urban LULC classes are united. The resulting projections act as a spatial constraint for the second stage, in which the full set of urban LULC classes is distributed. The resulting projections are better extrapolations of past spatial trends. Even after many iterations, the initially observed increased fragmentation of built-up areas is absent. This finding is confirmed when calculating the fractal dimension of the distributed built-up areas, which follow the observed trend within the base-year interval.

Although the automatic calibration can probably be further improved by improving the local search algorithm, as well as other aspects of the model, the initial outcomes are encouraging. Future applications of the model should include additional case study cities to investigate the robustness of the claims made in this study.

# 4. Urban growth projections

## 4.1 Introduction

Despite their current size, megacities like Dhaka and Beijing are still growing at a rapid pace with populations approaching 30 million in 2030 (UN, 2015). In addition, complete river deltas that hosting multiple large cities are rapidly urbanizing to become vast urban networks. This is the case in for instance the Pearl River delta where Guangzhou and Shenzhen act as mere hubs in a vast network of major urban agglomerations. In other areas, growth is slowing down or even stopped like in for instance the Tokyo-Yokohama

The topographic conditions, in which this urban development occurs, differ dramatically. One aspect is the position of cities within the river catchment. Deltaic cities like Dhaka and Calcutta for instance, face very different constraints from cities located more upstream like New Delhi or Lahore. The rugged terrain surrounding Tehran drives urban development towards more even landscape while the proximity to the Nile drives Cairo's growth pattern. Dhaka grows mostly along the north-south axis due to the adjacent rivers and wetlands while Beijing shows almost concentric development resulting in a very compact urban form. On a more local scale other factors determine urban development. These include for instance the proximity to main access roads, which often leads to ribbon development along major highways. Also locations in the vicinity of water bodies, public transport hubs and more importantly urban centres (including central business district) often determine the shape of cities. Obviously, these spatial relations change over time causing for instance reinforcement of local urban clusters, or in contrast dispersion to new areas. An extensive study of the characteristics of urban form has been performed by Batty & Longley (1994), which focusses on the fractal resemblance of the typical urbanisation patterns. Obviously, there are numerous other factors that ultimately determine shape of cities, including a range of socio-economic features that determine land prices, housing preferences, etc. Yet, in light of the goals of this study, what is important is is to acknowledge the differentiation in shapes, sizes and growth patterns between cities that determine to a large extent to what extent they are exposed to external hazards, including floods.

## 4.2 BAU for urban growth

The LULC change model by Veerbeek et al (2015) described in the previous chapter, has been used to develop growth projections for the 18 case study areas. As explained previously, these are based on a spatially explicit BAU-scenario in which transition rules that govern LULC changes have been derived from historic datasets. These are applied iteratively to develop future projections. So, for instance if a city's past urban growth consisted mainly of ribbon development, with the main road system as the attractor of new built-up areas, future developments will also concentrate along the infrastructural network. The focus on LULC implies that the representation of cities is based on 2d mapping. The expressiveness of the LULC maps is dependent on the remote sensing data (i.e. Landsat TM and ETM+ data) on which the LULC base maps are based. No further characterisation of land use (e.g. building types) has been applied. The representation of built-up areas has not been further extended by for instance population density attributes. This might be considered as a limitation, but on the other hand it keeps the base data and subsequent projection better manageable and omits a set of assumptions that are ultimately required to provide a more extensive characterisation of what is expressed in a single LULC cell.

To better characterise the assumptions the BAU-scenario is based up on, some of the key aspects are summarized below:

- *Decaying growth rate, stable growth*. The initial growth rate is based on the observed rural-urban and intra-urban LULC-transitions in the interval used for the model calibration. Typically this covers a 20 year period, between 1990 and 2010. Since the growth rate acts as an operator on a finite (and effectively declining) number of candidate cells for LULC transitions, the growth rate decays over time. This results in a stable urban growth and prevents exponential growth (see chapter 3).

- *Fixed weights and transition rules*. The derived weights and subsequent LULC transition rules are fixed over time. This means that the weights used to calculate the transition probabilities (e.g. a high probability for urban development close to existing patches of urbanisation) remain unaltered during the future projections. No top-down interventions that might alter these weights are applied. This is an essential feature of a BAU scenario.

- *Horizon 2060 with 5 year intervals*. As explained in chapter 3, the projections are iteratively developed using 5 year intervals with a set limit of 2060. With 2010 as a base year for the projections, this year represents a 50 year horizon. The choice of a 50 year period is somewhat arbitrary but represents a period between the typical long term horizon for urban planning of about 20 years and the customary horizon of climate change projections of 50 to 100 years;

From a model perspective, the horizon for these scenarios is arbitrary; projections can be developed for 2100 or beyond by simply increasing the number of iterations. Yet, an argument can be made concerning the reliability of the projections since the typical temporal interval used for model calibration, spanned 20 years or less. The extrapolated trends over a projection period that well exceeds that 20 year period, might be judged as disproportionately large. On the other hand, this qualification might be caused by misinterpreting projections for predictions. Again, the aim of the projections is to simply explore what would happen if current trends in urban growth would continue into the future. Especially in developing countries, the volatility in socio-economic development in combination with changing policies and demographics include too many uncertainties to make robust long term future predictions. The relatively long projection period is merely chosen to ensure that the differentiation in growth characteristics becomes explicit. Veerbeek (2015) claims that the produced growth scenario for Beijing for instance, is nothing more than an exploration of a 'what-if' scenario where the consequences of an observed urban growth trend for the city of Beijing, are projected into the future. The subsequent scenario can therefore be regarded as a 'business-as-usual' scenario which is a common approach in many disciplines. As such, the projections can be considered as baseline scenarios from which alternatives can be developed and evaluated.

With these assumptions and considerations, urban growth scenarios have been developed for the 18 megacities. The resulting growth patterns are presented in detail in Appendix A3, where for each city the current as well as the projected new development is presented in a series of maps. An more in-depth coverage of growth analysis is presented in the following sections.

## 4.3 Historic and projected urban growth

When comparing the respective urban footprints, the sheer magnitude of the observed as well as projected growth becomes clear. In 1990 for instance, the combined Guangzhou-Shenzhen urban area consisted of about 815 km2. The urban footprint of this network of cities ranked in 3rd place after 'giants' like Mexico City and Beijing, for which the footprint was almost double the size (1535 and 1472 km2 respectively). In only 25 years things turned around: currently the Guangzhou-Shenzhen urban footprint is more than double the size of the Mexico City and Beijing urban footprints. The exploding growth of Guangzhou-Shenzhen is symptomatic for all Chinese cities in this research: in 2060 Guangzhou-Shenzhen, Beijing and Shanghai are projected to have the largest urban footprint of all rapid growing megacities. The growth is illustrated in Figure 13, where the estimated urban footprint for 1990 and 2010 is depicted, as well as the projected urban growth between the base year 2010 and the 2060 horizon. The footprint for 1990 and 2010 is based on the LULC classification (see Appendix A1) while the projected growth underwent additional post-processing to distinguish different

types of urban growth (i.e. infill, expansion and leapfrogging development).
Many of the key statistics illustrate the extraordinary transition this area is experiencing:

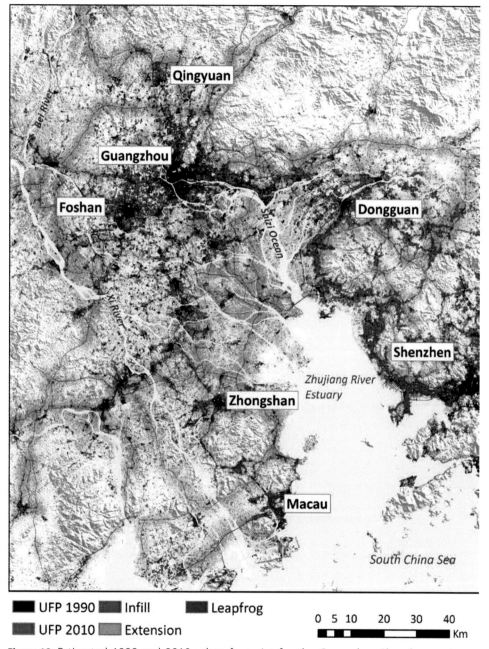

**Figure 13:** Estimated 1990 and 2010 urban footprint for the Guangzhou-Shenzhen region as well as the projected urban growth between 2010-2060.

- *Urban footprint:* an urban footprint that is projected to grow more than 10-fold in size in the period 1990-2060;
- *Growth rate.* Growth rates that range from an estimated initial 10.1% in 1990 to 0.8% in 2060;
- Density. A predominantly suburban and rural footprint (71%) in 1990 that transforms towards 2060 into a high-density network city dominated by urbanised built-up areas (68%);
- *Ribbon development.* Extensive urban development along the major corridors that connect the regional centres: Dongguan, Foshan, Guangzhou, Macau, Qingyuan, Shenzhen and Zhongshan;
- *'Flood insensitive'-growth.* Extensive development of the coastal zones (e.g. west and northwest of Shenzhen), around the coastal wetlands (e.g. at the Shizi Ocean) and along stretches of the major rivers (e.g. on the Xi-riverbank southwest of Foshan);
- *Growth type.* The projected development consist predominantly of urban extension (68%), which enlarges existing patches of contiguous built-up areas;
- *Distribution.* Increasing fragmentation of the perimeter of the urban footprint but an decrease of open areas in the urban cores.

While the growth statistics in many of the metropolitan areas in this study might be exceptional, analysis of the past and projected urban growth show considerable dif-

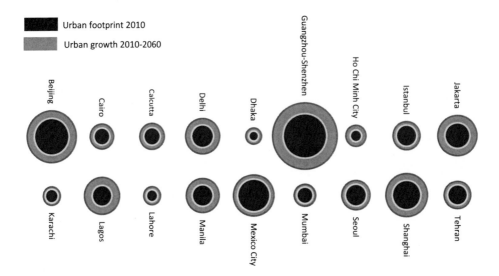

**Figure 14:** Estimated urban footprint and growth

ferences in size, shape and growth characteristics between the different case study megacities. In 2010 for instance, the urban footprint of the Mexico City metropolitan area was estimated at 2209 km2. In contrast, for Cairo the footprint only covered about 604.9 km2. Compared to the estimated growth, Cairo's footprint is tiny when set against the projected growth of Jakarta's urban footprint, which is expected to increase by 1948 km2 in a period 50 years.

To first of all, provide better sense of the relative size differences between the respective urban footprints well as the magnitude of the projected growth, they are represented in a bubble chart (see Figure 14). From the chart, the relatively large urban footprints of Guangzhou-Shenzhen which outsizes all other metropolitan areas, is somewhat expected. What seems more remarkable are the small footprints of Dhaka, Ho Chi Minh City, Karachi and Lahore that nevertheless are all classified as megacities (albeit with significantly lower populations). Compared to the mean urban footprint, which is about 1160 km2, these cities have small footprints of 390 km2 or less. This provides a clue for the extreme population densities that can be found in these cities. In Karachi for instance, the estimated mean population density in 2015 was about 24100/km2 (Demographia, 2016). In all three Chinese cities the densities were only a fraction of this. In Beijing for example, the population density was more than 4 times lower at 5200/km2 (ibid).

The size of the urban footprint might provide a clue about the dominant urban growth type. Typically, urban extension and particularly leapfrogging development increase the footprint into a vast but fragmented urban landscape. Development dominated by urban infill on the other hand keeps the perimeter of the urban footprint intact but creates a single contiguous compact built-up area. This would suggest that cities with disproportionately large projected urban footprints would also exhibit the corresponding growth types. Yet, while the proportion of leapfrogging development is indeed high for Guangzhou-Shenzhen and Shanghai (23% and 34% respectively) Beijing's concentric development is largely due to extension of the existing main urban core. Furthermore, leapfrogging development is also an important contributor to urban growth in Calcutta, Delhi and Jakarta which do all exhibit more moderately sized urban footprints.

Densification due to infill is considerable in Mexico City, Mumbai and Istanbul but never amounts to more than 19% of the total projected growth. In fact, for all cities in this study, urban extension is the predominant growth type. On average urban extension amounts for 74% of the total urban expansion while the contribution of infill and leapfrogging development is limited to 11% and 15% respectively. This is partially due to the rigid definition of urban extension, where the existence of a contiguous built-up area can depend on a single 'strand' of cells with an urban LULC classification. Yet, the ample fraction of urban extension can also be explained from a 'spatial' perspective:

- The available area this is suitable for infill is only a fraction of the space available for urban extension or leapfrogging development. Given the fact that over the course of 50 years, most of the cities in this study double in size, the fraction of open land patches is simply too small to accommodate the projected urban growth.
- In most cases, the proximity of existing urban built-up areas is a requirement or at least strong driver for new development. On a fundamental level this relates to the essence of urban areas: a concentration of built-up areas. Thus, the probability that new built-up areas appear adjacent to existing areas is in most cases larger than disjoint development.

The contribution of the infill, extension and leapfrogging to the projected development between 2010 and 2060, is included in the base statistics for each city in Appendix A3. Inspection of these fractions reinforce the notion that urban growth is a complex phenomenon; simple growth trends cannot be observed and further exploration of the growth characteristics is required to gain more insight in the growth behaviour of the cities.

The estimated growth rates based on the observed LULC changes between the base years 1990 and 2010, are presented in Table 5 as doubling periods, i.e. the estimated periods of time required for the urban footprints to double in size. For instance the urban footprint of the Guangzhou-Shenzhen region of 1990 requires only 8.3 years to

**Table 5:** Doubling periods

| City | Doubling [y] 1990/2010 | City | Doubling [y] 1990/2010 |
|------|------------------------|------|------------------------|
| Beijing | 21.8/48.3 | Karachi | 21.4/54.8 |
| Cairo | 23.0/46.6 | Lagos | 10.4/30.4 |
| Calcutta | 27.3/45.1 | Lahore | 16.9/43.1 |
| Delhi | 19.6/39.7 | Manila | 14.3/38.1 |
| Dhaka | 18.5/35.9 | Mexico City | 37.8/76.9 |
| Guangzhou-Shenzhen | 8.3/43.5 | Mumbai | 39.8/62.0 |
| Ho Chi Minh City | 10.9/35.7 | Seoul | 31.3/60.3 |
| Istanbul | 24.7/62.2 | Shanghai | 13.7/45.6 |
| Jakarta | 13.7/38.4 | Tehran | 16.2/59.0 |
| *Mean* | *20.5/48.1* | | |

double in size. In contrast, this would take an estimated 39.8 years for Mumbai. These differences sets the extraordinary rapid development that took place in the Pearl River delta over the past decades against the more moderate growth of the already more established megacity of Mumbai. On average the doubling time is estimated at around 20.5 years. Apart from Guangzhou-Shenzhen, also Ho Chi Minh City and Lagos show extreme growth with estimated doubling times of about 10 years. Mexico City and Seoul on the other hand show doubling times exceeding 30 years. Intuitively, the doubling times seem strongly related to the size of the urban footprint in 1990, which serves as the initial base year. Consequently, smaller cities can quickly double in size, while for larger cities this takes considerable longer. Yet, after inspection of Figure 14, this rule is not necessarily always valid: Although somewhat smaller, the urban footprint of Ho Chi Minh City is not dwarfed by the footprint of Mumbai. In fact, the 1990 footprint of the rapidly growing Lagos is significantly larger than for instance the footprint of Karachi. Yet, the doubling time for Karachi is estimated at 21.4 years, which is more than twice as long as it takes for Lagos to double in size.

The impact of size on the doubling period does seem strong when comparing the periods derived from the projected growth in the 2010-2060 interval to the observed growth between 1990 and 2010. The periods are significantly longer due to the absence of exponential urban growth that would be required to sustain the historic growth rates. Also here, the differentiation between cities is significant. Istanbul for instance, showed an initial doubling period of 24.7 years which is slightly above the mean. Yet, in the projected interval 2010-2060 the doubling period increases to 62.2 years which is the second longest of all cities. Calcutta shows the opposite behaviour: the historic doubling period is higher than average (27.3 years) while the future period is lower than average (45.1).

## 4.3.1 Urban composition

The proportion of infill, extension and leapfrogging development causes the urban footprint to change over time. In many of the cities in this study, this implies an increasing densification illustrated by a shift from suburban into urban built-up areas. Typically, small villages as well as low density historic development are converted into new residential districts with significantly higher densities. This is illustrated in the shifting fractions of urban landscape classes in the urban footprint. These classes are based on metrics developed by Angel et al (2007) that were applied to the urban growth projections. The classification describes the respective LULC distributions into a set of six urban landscape classes that are based on the concept of *urbanness* (ibid), which is defined as:

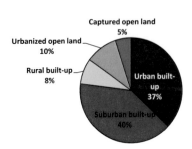 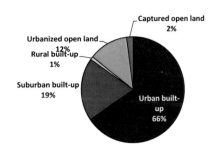

**Figure 15:** Urban landscape composition for Guangzhou-Shenzhen in 2010 (left) and 2060 (right)

*The percent of built-up area in a circular neighbourhood of 564m (~km2) centred on the raster cell of interest.*

With this concept, the following urban landscape classes are defined and extended with subsequent metrics:

1. *Urban built-up.* Built-up pixels with urbanness values equal or greater than 50%;
2. *Suburban built-up.* Built-up pixels with urbanness values between 10-50%;
3. *Rural built-up*: Built-up pixels with urbanness values less or equal than 10%;
4. *Urbanized open land.* Undeveloped land with urbanness values greater than 50%;
5. *Captured open land.* Patches (i.e. sets of adjacent raster cells) of undeveloped land, less than 200 hectares, that are completely surrounded by the urban built-up, suburban built-up, and urbanized open land raster cells;
6. *Rural open land*: Undeveloped land not classified as urbanized open land or captured open land.

Using these metrics, the changes in urban footprint composition are illustrated in Figure 15, for the already earlier presented example of the Guangzhou-Shenzhen metropolitan area.

The significant reduction of the proportion of suburban built-up areas as well as the doubling of the fraction of urban built-up areas illustrates the changing character of the region. On top of that, the transformation of the region also marks the decline of villages and smaller towns, which are eradicated beyond 2060. In contrast, in 2010 small settlements still contributed about 8% to the urban footprint. The open land fractions show opposing trends: the fraction of urbanised open land which typically indicates undeveloped land along the fringes increases while the fraction of captured open land, which identities small 'holes' (e.g. parks) within built-up areas is decreasing due to future infill. Inspection of the actual maps (Figure 13 and Appendix A4.6) as well as the actual development of the landscape classes (Appendix A4.6, Fig. 59) further illustrate these transformations. The most prominent outcome though are summarised below:

- Densification. The urban footprint of many of the case study cities initially shows a significant portion of suburban and rural built-up areas, e.g. in Dhaka (2005): 24% urban, 46% suburban, 23% rural. In 2060 the ratios have shifted to 69% urban, 19% suburban and 2% rural. Significant but less radical shifts towards higher densities are projected for Beijing, Delhi and Tehran.
- Saturation. Extensively urbanised urban footprints in which also the remaining captured open land is transformed into urban built-up area, creating a vast contiguous urban core. This is for instance the case in Karachi, where the fraction of urban built-up area was already estimated at 64% in 2005 and is projected to further increase to 79% in 2060. Other cities with similar projected trends are Beijing, Istanbul, Lagos and Mexico City with estimated fractions of urban built-up areas of around 80% in 2060.
- Disappearance of villages. Apart from Guangzhou-Shenzhen, a significant reduction of the fraction of rural built-up areas is projected in Calcutta: 23% (2005) to 9% (2060) and in Dhaka: 13% (2000) to 2% (2060). Typically, these villages are absorbed into the suburban areas at the extending fringes of the urban cores. Leapfrogging development typically only comprises a small fraction the projected urban growth, which limits the appearance of new settlements in the vicinity.
- Sprawl. In a few megacities, the ratio of extensive suburban built-up areas remains relatively stable over the projection period. A prime example is Shanghai, where the urban footprint in 2010 consisted for more than half of suburbs (56%). In 2060, this fraction is expected to decline only slightly declines to 52%. Also in Calcutta and Mumbai this is the case with values ranging over the projection period from 41%-42% and 49%-42% respectively.

## 4.4 Growth potential and characteristics

### 4.4.1 Spatial constraints

While high urban growth rates feed the appetite of cities for territorial expansion, the degree-of-freedom for urban growth is in many cases limited by geographic features that make occupation cumbersome if not impossible. Mumbai for instance is located on a landfilled peninsula which is almost completely saturated by dense built-up areas, virtually pushing the city into the sea (Banerjee-Guha, 2002; DNA India, 2017). Facing similar challenges, Jakarta undertakes and extensive land reclamation project to provide space for new residential areas combined with a barrier to improve the city's coastal protection (Ministry for Economic Affairs, 2015).

In most of the metropolitan case study areas in this study, the geographic features of the territories have big implications on the urbanization patterns. Rugged mountains, extensive wetlands but mostly the actual shape of the shorelines defines the boundaries for the quest for space. In Table 6 the dominant growth constraining geographic features are indicated for all 17 case study areas. These can be observed in more detail in Appendix A3, that provides extensive maps of for each city.

Cities that are particularly constrained are Seoul, where the rugged mountainous area only provides space in the narrow valleys or in the coastal wetlands of Incheon as well as the already mentioned metropolitan area of Mumbai. Also Istanbul faces a limited degree-of-freedom since its hinterland is on both the eastern and western sides constrained by rough mountain terrain. Space for urban expansion is therefore mainly pro-

**Table 6:** Dominant geographic growth constraints (indicated in black) and growth statistics

| City | Geographic feature | | |
|---|---|---|---|
| | Coast line | Mountains | River delta |
| Beijing | | ■ | |
| Cairo | | | ■ |
| Calcutta | | | ■ |
| Dhaka | | | ■ |
| Guangzhou-Shenzhen | ■ | ■ | ■ |
| Ho Chi Minh City | | | ■ |
| Istanbul | ■ | ■ | |
| Jakarta | ■ | | |
| Karachi | ■ | | |
| Lagos | ■ | | |
| Lahore | | | |
| Manila | ■ | | |
| Mexico City | | ■ | |
| Mumbai | ■ | ■ | |
| Seoul | ■ | ■ | |
| Shanghai | ■ | | ■ |
| Tehran | | ■ | |

vided along the coastline (Appendix 3.8) towards the suburbs of Esenyurt and Gebze.

## 4.4.2 Growth attractors

Extensive concentric urban growth, as explained by central place theory (e.g. Christaller et al, 1966; McCann, 2012) only occurred in Beijing where the Forbidden City acts as the geographic centre from which the city expanded outward. Although Beijing's territory is bounded by the Xishan and Yanshan mountain ranges in the north and east, there is ample space for further expansion as was illustrated in the previous chapter or in Appendix A3.1. This sets Beijing apart from many of the other cities in this study, that are located in river deltas and/or coastal zones thus facing a limited degree-of-freedom for future urban expansion. A good example is Manila, for which the growth projections predominantly show an expansion of the primary urban cluster. Yet the city is constrained by two bays on the east and west, which "squeezes" urban growth along the north-south axis (Appendix 3.13).

Cities where growth occurs mainly around secondary clusters are Delhi, Dhaka, Guangzhou-Shenzhen, Lagos, Mexico City and Mumbai. This means that growth is not only occurring around the main urban cluster, which often indicates the central business district, but also concentrates around other urban nuclei. This is mostly apparent in Delhi (Appendix A3.4), where the adjacent cities of Faridabad, Greater Noida, Gurgaon and Bulandshahr serve as important regional clusters within the Delhi metropolitan area. This behaviour seems not driven by landscape features, i.e. Delhi is not surrounded by rugged mountains or extensive rivers and wetlands. Faridabad for instance acts as a regional industrial centre and therefore feeds on rapid rural-urban migration (Acharya et al, 2016). This phenomenon was already occurring in the 1995-2010 interval and is therefore further extrapolated in the growth scenario. In the five other cities, multi-nuclei development is dictated by landscape features. The ring of mountains surrounding Mexico City for instance causes extensive growth of adjacent cities like Cuernavaca and Toluca, that are disjoint from the primary urban cluster that hosts Mexico City's central business district. At the same time, the constrained growth possibilities for the main urban cluster results in infill of the last remaining patches of open land. In the growth scenario for Dhaka, the surrounding rivers and wetlands cause extensive growth in the northern (e.g. Tongi and Joydebpur), western (Savar) and southern (Narayanganj) regional clusters. While the regional clusters in the Dhaka metropolitan area distributed over a relatively small area, in Guangzhou-Shenzhen the network of regional clusters covers many thousands of km². A few decades ago, the respective clusters could be regarded as isolated cities and towns that all served as regional centers. The massive urbanisation of the Pearl river delta though, forces a new perspective in which the cluster act local centres within a vast urban network bounded by sea and mountains. The road network, which has been developed over the past decades, serves as a linear 'connector' between the clusters; extensive ribbon development is projected along these infrastructural corridors. Over the past decades this trend already started between Guangzhou-Dongguan-Shenzhen and ultimately Hong Kong. This eastern corridor might be extended along the western coast of the South China Sea, stretching from Guangzhou to Macau.

Other cities where ribbon development along major roads is seems the dominant urban growth pattern are Ho Chi Minh City, Lahore and Tehran. Although for most cities, small scale ribbon development occurs (e.g. north of Cairo), in Lahore for instance the major highways drive the dendrite-like urban development outwords of the city, connecting suburbs like Sargodha, Gujranwala or Bhai Pheru to Lahore city. Major highways in the Tehran metropolitan area connect the city to suburbs like Eslamshahr, Karaj, Mamazand and Varamin. Also here, extensive ribbon development is projected causing the city to develop built-up areas along these four major axes. Finally, for Ho Chi Minh City, extensive ribbon development is projected which will be further illustrated from the actual weight distribution derived from the urban growth model.

For each city, the identified push and pull-factors can be traced back to the derived weights of the calibrated growth models that determine the probability for LULC transitions (see chapter 3). Since these consist of aggregate probabilities which are based on the combined probabilities of individual factors (see equation 6), the identification

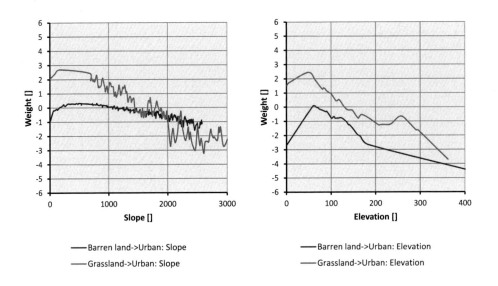

**Figure 16:** Derived weights distribution of slope and elevation for Seoul

of dominant push- or pull factors is not always straightforward. For instance, the influence of the road network as an attractor of urban growth is based on the combined weights for the distance to urban roads, rural roads, public transport hubs as well as the road density. Yet, in a given cell also all other factors exert their positive and negative weights, which ultimately determine the rank for a given transition. Nevertheless, in some cases the derived weights show a clear relation. This is illustrated in Figure 16 for the influence of terrain characteristics in Seoul and the for the road network in Ho Chi Minh City in Figure 17. The influence of terrain is subdivided into slope (Figure 16, left) and elevation (Figure 16, right) for two transitions to high density built-up areas: from grassland and from barren land. Infrastructure is subdivided into distance to main roads, secondary roads and public transport for the transitions to high density built-up areas from grassland (Figure 17, left) and barren land (Figure 17, right).

Note that the increments on the horizontal axis do not necessarily represent real units (e.g. metres) but are based on an inverse distance relation translated into categorical map values which represent unequal increments. This is done to give more importance to lower values instead of extreme values, thus better representing local phenomena. For the transition from grassland, Figure 17 clearly illustrates the transition from pull into push factors; for increasing values for slope and elevation the weights decrease and turn from a positive sign (i.e. pull) to a negative sign (push). In other word, higher values for slope and elevation have a negative impact on the transition probability. With the exception of lower slope values, where the effect is almost neutral, the de-

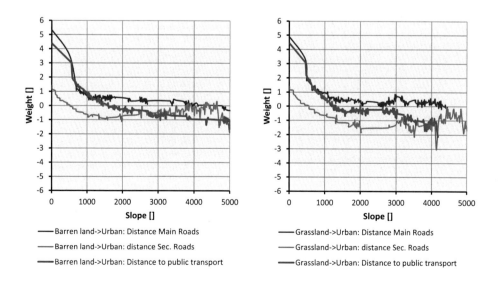

**Figure 17:** Derived weights distribution of distance to infrastructure for Ho Chi Minh City

rived weights for the transition from barren land are predominantly negative.

The decaying effect as well as the transition from pull- to push factor can be more clearly observed in Figure 17, where the close proximity to especially main roads and public transport is expressed in high positive weight values. This impact declines quickly for increasing values, which seems intuitive in Ho Chi Minh City where high density urban areas are located around major infrastructure (Appendix A4.7). The impact of public transport and secondary roads seems very much spatially bounded: only within the immediate vicinity the weights exert a pulling effect after which the impact quickly turns neutral or tips over into a pushing effect.

With the exception of Istanbul, all cities in this assessment are located in the vicinity of major rivers, only a few cities the rivers seems the driving force for the formation of new built-up areas. Only Cairo and Calcutta have urban forms that stretch along the Nile and Hooghly River respectively. In Cairo extensive urban development along the Nile River is projected south of the current urban clusters in the direction of Helwan which is currently a suburb of Cairo, virtually disconnected from the urban footprint (Appendix A3.2). Calcutta's projected urban development on the other hand is only partially following the river contour, mainly reinforcing fragmented stretches of built-up areas in the northern direction of the city. More significant development though is projected adjacent to the extensive wetland area east of Calcutta. This region that borders with Bangladesh develops rapidly to become a vast, yet fragmented urban region with extensive development along the riverbanks of the Ichamati and Kalindi River (Appendix A3.3).

## 4.5 Conclusions

The presented growth analysis has not converged into a few straightforward general principles that can be transferred to other cases. Urban growth is differentiated and not easily captured in a single driver or predictable growth pattern. In a way, this observation is consistent with the modelling approach, in which a generic model is adapted for each case separately. The unique combination of derived push- and pull factors (i.e. weights), initialized by a region specific base map in combination with a set of thematic maps, produced spatiotemporal unique complex behaviour. This ultimately means that the produced growth projections should be analysed as separate cases. The analysis provided in this chapter merely described the observed range in some of the most important growth aspects like urban footprint size and composition, growth rate, spatial constraints as well as growth attractors. The most prominent outcomes of this analysis were:

- The Chinese cities provided the upper bound for the observed and projected growth statistics; i.e. they combined extensive urban footprints with rapid urban development. This culminated into the largest metropolitan areas in this study. Yet, the way that growth manifested itself differs considerably. Growth in Beijing was characterised by concentric outward development while Shanghai's grows expanded into a vast network of fragmented sprawl.
- The only common factor for all growth projections was that the majority of urban growth consisted of urban extension. The fractions for other growth types like infill or leapfrogging development were considerably lower. As a consequence, the urban footprint of all cities was expanding. This could be seen in the doubling periods: for 13 out of 18 cities, the urban footprint is projected to double in size in the 2010-2060 interval.
- Apart from Istanbul, Mexico City and Seoul all cities had sufficient buffer space to absorb the projected growth. Geographic features like coastlines, river deltas and mountainous areas rather directed the growth than were limiting it.
- While for each city a unique set of push and pull-factors determined the growth pattern, the weights derived from the proximity to surface water or streams was negligible. In other words, no evidence was found that urban growth is affected by the city's orientation to surface water.

The presented outcomes would obviously be very different if a deliberate choice would have been made to adjust the growth parameters. The relatively narrow constraints that were used for the development of the BAU-scenario are due to data limitations but were also intended. Although available, population and economic data and associated socioeconomic scenarios that might alter growth rates and weights-of-evidence were intentionally omitted. This was a deliberate choice to ensure that the outcomes are as consistent as possible and not subject to a potential bias due to interpretation of additional data or scenarios.

# 5. Future riverine flooding in megacities

## 5.1 Introduction

The growth of megacities gave rise to an unprecedented concentration of people and assets on limited territories which are in many cases exposed to a multitude of natural hazards including earthquakes, typhoons, forest fires and not in the least: floods. In fact over the last decades, floods are responsible for the majority of impacts from natural hazards (Jongman et al, 2012). These impacts will exacerbate in the future as a consequence of climate change, illustrated in a large number studies that estimate future flood impacts as a result of sea level rise, more intense storm surges, precipitation and often associated river discharges.

The outcomes of large comparative studies are often presented as a ranking of cities most exposed to future flooding sometimes in combination with the expected growth they undergo. While such lists are fairly common for sea level rise and associated coastal flooding, a ranking for rapid growing cities exposed to riverine flooding has thus far not been produced. This might be due to the fact that simple extrapolations of historic flood impacts are less likely candidates for obtaining approximations of future growth induced estimates. Since the location of built-up areas along rivers shows a large variability across cities, only a spatially explicit assessment can produce figures that provide some level of confidence. To perform such a study, two requirements need to be fulfilled: i) the availability of a set of spatially explicit urban growth scenarios for a substantial number of flood prone metropolitan areas and ii) a set of flood inundation maps that cover the urban extent associated to those cities. The development of a urban growth scenarios for the purpose of flood risk assessment was an important ingredient in a former European Union-funded FP7 project: Collaborative Research on Flood Resilience in Urban areas (Djordjević et al, 2011), in which a spatially explicit urban growth model has been developed to produce growth projections for four rapid growing Asian megacities. The outcomes of this model have been extended to develop growth scenarios for 14 additional megacities across the globe. These were complemented by a set of inundation maps with global coverage produced by the Global Flood Risk with IMAGE Scenarios (GLOFRIS) model which include a set of river flood scenarios with wide range of return periods (Winsemius et al, 2013).

In this section, the outcomes are presented from a study that attempt to assess the influence of urban growth on the exposure to riverine flooding by combining spatially explicit urban growth scenarios for 18 rapidly developing megacities with inundation maps produced by floods with return periods ranging between 10 and 1000 years. The outcomes produce an alternative ranking of cities most exposed to future flooding, by assessing the size of the projected urban footprint located in the estimated floodplains. More importantly though, the assessment also focuses on the proportionality of the urban flood risk development, by analysing which fraction of the projected development is located within the floodplains and more importantly if that fraction increases, remains stable or diminishes. The resulting classification indicates for which cities future urban development has a disproportionate effect on future flood risk and provides a case for flood sensitive urban planning policies. Finally, by looking at the changes in growth induced flood risk, the study attempts to provide a better insight in quantitative contribution of urban growth to flood risk. The resulting baseline projections might be a useful tool for assessing the impact of flood risk management policies, strategies or measures that integrate controlled urban development.

In this chapter first an introduction is made to models and scenarios used in this study. This is followed by a description of the used datasets including the inundation maps produced in GLOFRIS. Since the datasets used for this study differ significantly in level of detail, a separate section is dedicated to the validation of the procedures and outcomes. The main body of the chapter is reserved for the produced results, the subsequent interpretation and ultimately the classification and ranking of the cities based on their growth induced estimated future flood exposure. Due to the large amount of produced data, associated graphs and maps a major part of this study is located in a separate Appendix A3 that acts as a "flood atlas". In this Appendix, the results for individual cities are presented including the key-statistics that were used in their evaluation.

The bulk of material presented here has previously been presented in a study commissioned by the Netherlands Environment Assessment Agency (Veerbeek et al, 2016). The initial focus of the study was to assess the usability of the flood maps produced by a global river flood model with urban growth scenarios with a considerably higher level of detail.

## 5.2 Urban growth and floods

Flooding accounts for the majority impacts of all natural hazards and is expected to significantly increase in intensity and variability in the future due to the impacts of climate change (e.g. Kundzewicz et al, 2013). This is especially the case in rapidly urbanizing deltas where the flood hazard as well as the exposure and vulnerability of an increasing concentration of people and assets is growing. This rapid urbanisation does not only increase the susceptibility to flooding but in many cases also acts as an active component that changes the microclimate and regional precipitation locally (e.g. Pathirana et al, 2014). This is especially the case in vast metropolitan areas covering areas of 100s of square kilometres and hosting often several tens of millions of inhabitants.

Although many models exist to develop flood impact estimations, the scale of megacities still proves to be a challenge concept. On the one hand models exist that can successfully be applied at lower scales in order to express the differentiation required in cities, the sheer size of megacities and associated metropolitan areas often proves intractable. Yet, models used for regional flood impact assessments are often too coarse and lack detailed integration of spatial characteristics that ultimately determine the produced outcomes. Apart from the required level of detail in the flood exposed urban areas, this issue also holds for the flood models from which the inundation maps are produced.

In a selected number of countries, the production of flood inundations maps for different exceedance probabilities is common practise. To accelerate this process in Europe, the EU-flood directive for instance dictates that for all major rivers flood hazard maps are produced for at least the 100Y event. In the developing world, where the major portion of urban growth is located, the availability of flood inundation maps is often limited. Flood maps are often only available for actual flood events which limit their application for comparative assessments where a more uniform approach is required. Furthermore, only in a few cases inundations maps associated to different return periods are available which limits assessments to a single event. This is for instance the case in a study performed on Dhaka (Khan et al, 2015) where the inundation map associated to the 2004 flood is projected on an urban growth scenario for 2050. Such case-based approaches are illustrative for many studies and prevent wider application to larger comparative studies focussing on the nexus of riverine flooding and urban development. Yet, while many large metropolitan areas in the world are located along coastlines, more often so they are located in along major rivers. With the exception of impacts caused by major catastrophic events like the Tsunami's in East and Southeast Asia (e.g. Suppasri et al, 2011; Leone et al, 2011), the impacts of riverine flooding are the dominating contributor in the annual global flood impacts (Jongman et al, 2012). This was again illustrated during the recent European floods in May and July 2016, causing casualties and significant damages in Germany and France (e.g. BBC, 2016) where many rivers and tributaries exceeded their peak discharge capacities.

With many cities rapidly expanding within the flood extent of the world's rivers, the need for a comparative study focussing on the effects of riverine flooding on urban and especially metropolitan regions only becomes more pressing. One of the components that greatly simplifies such comparisons is the GLOFRIS dataset, which provides inundation maps based on a set of riverine flood events with a range of return periods with global coverage (Winsemius et al, 2013). This means that the dataset provides a basis for assessments ranging from relatively frequent events with return periods of 10 years to extremely rare events that are estimated to happen only once every 1000 years. The uniform approach in the underlying model ensures a relatively homogenous dataset from which tiles can be extracted for use in individual regional case studies.

## 5.2.1 Datasets

The GLOFRIS dataset provides return periods of 1, 10, 25, 50, 100, 250, 500 and 1000 years, which enables an assessment beyond the 100Y event that is often used in analysis. Using multiple return periods allows for a characterisation flood extent (and ultimately the flood risk) as a function of different return periods. This concerns questions about for instance the proportionality of the flood extent (e.g De Bruin, 2004) for increasing return periods. This might significantly differ between cities, where in some cases the flood extent will gradually increase for less frequent floods while other cities experience sudden stepwise changes in the flood extent once a threshold is exceeded. For this study the decision was made to omit the 1 year flood events. In general, these are associated to relatively small inundation depths and a limited flood extent. In comparison to the relative coarse resolution of the GLOFRIS dataset, inclusion of the 1 year events would seem inconsistent; due to the high frequency, irregularities stemming from the coarse resolution would have a disproportionate effect on the overall outcomes.

The produced urban growth scenarios all start from the second base year used for the training of the LULC-change models and cover a period till 2060 using 5 year increments. In most cases this covers a period of 50 years since the the projections are made from 2010 on. Since the base maps for the models are derived from Landsat datasets, the cell size of the LULC maps is 30m. The extent differs per city and is based on an approximation of the relevant metropolitan region including an additional buffer zone to accommodate the expected future growth. In practise this results in very differently sized map extents, ranging from about 2.9 million cells for Dhaka (representing about 2622 km2) to about 43.5 million cells for Shanghai (representing 39138 km2).

The combination of LULC map instances for the growth scenarios with the earlier described range of return periods results in 56 (8 future projections, 7 return periods) individual flood assessment instances per city.

Classification has been performed using a maximum-likelihood based supervised classification combined with a multi-temporal classification (Bruzzone and Serpico, 1997) in order to limit classification errors. For the LULC classification, the NLCD 2001 Land

Cover Class Definitions (Fry, 2013) have been used. One of the main features of this classification scheme is that urban areas are subdivided into 4 different classes: open as well as low, medium and high intensity developed areas. This subdivision provides a satisfactory trade-off between on the one hand the required level of expressiveness of differentiation in urban densities and on the other hand, a sufficient level of abstraction that fits with the limited level of information that can be extracted from the Landsat based datasets from which the LULC maps are derived.

Additional error corrections were made by comparing Google Earth™ multi-temporal imagery to the produced LULC maps. Although this process is relatively straightforward, it is time consuming and still depends largely on the quality of the initial Landsat maps for the different regions. For Jakarta, for instance, it is almost impossible to obtain base maps that are (virtually) cloud free. This means that the multi-temporal classification process becomes differentiated across different regions of the basemaps; in some regions years have to be discarded since they suffer from substantial cloud cover while for others more instances are available since they are cloud free within the range used for classification.

Ultimately, the cloud-cover affected quality differences between the available Landsat datasets lead to some adjustments of the standard interval used for the base years: while for most cities the initial base years were set at 1990 and 2010, for quite some cities these had to adjusted (e.g. 1995 and 2010), thus shortening the period from which growth extrapolations were made.

The derived LULC classification were further adjusted by application of an urban landscape analysis (Angel et al, 2007) from which additional urban features were derived, e.g. urbanised and captured open land, rural built-up areas, etc.). This feature extraction is based on the characteristics of the urban patterns in the LULC maps instead of individual cells in the original Landsat datasets. For instance, green urban areas (e.g. park) are in the initial pixel-based classification scheme classified as grassland or shrubs. By discovering that they green zone is surrounded by built-up area, the feature might be identified as urbanised open land which differentiates it from for instance pastures in rural area. Thus, urban landscape analysis provides a complementary evaluation and subsequent characterisation of a derived LULC cover that extents the initial classification definitions. An overview of the classification definitions can be found in Appendix A1.

The analysis of urban growth and the associated distribution of urban areas (i.e. the shape or form of the city) is a discipline in its own right. Although a detailed description of the methods and metrics to characterise urban form are beyond the scope of this study, some are vital for a uniform assessment and comparison between the cities. The most prominent issue for a fair comparison is the geographic extent on which analysis is based on. The different tile size (i.e. map extent) used for the development scenarios to some extent reflect the size of the cities and their expected growth contours. Yet, especially in more fragmented metropolitan area, choosing a proper extent for modelling as well as analysis is not necessarily straightforward but could substantially influence many of the key statistics required to characterise growth, urban contour and ultimately the size of urban area exposed to flooding, which is the prime motive of this study. To overcome this shortcoming, analysis is performed only within the urban footprint of the area. This concept is operationalized by using metric developed by Angel et al (2007), which provides a set of analysis criteria to classify built-up areas. These criteria focus on pattern analysis of raster cells rather than on qualitative criteria.

The issue of a somewhat arbitrary choice of boundary for modelling and analysis is very well exemplified by the Guangzhou-Shenzhen region in the Pearl River delta. The consists of a vast network of urban agglomerations, consisting of 9 different cities: Shenzhen, Dongguan, Huizhou, Zhuhai, Zhongshan, Jiangmen, Guangzhou, Foshan, and Zhaoqing. Currently complete area hosts about 57 million inhabitants, which makes it with this definition the largest metropolitan area of the world. In terms of population distribution, the area does not have a city that outranks all other cities in the area. Thus, for this study the area has been taken as a single urban agglomeration which might skew some of the results. Yet, many of the metropolitan areas in this study consist of multiple urban agglomerations. The only difference in these cases, a single city (e.g. Beijing) acts as the overall centre, dominating the area in terms of size, population as well as urban functions and facilities.

Although the produced growth scenarios are extrapolations of past growth trends, this doesn't mean that observed trends in LULC-changes are a continuation of the same pattern formations that occurred during the interval between base years. The actual envelope which spatial trends can propagate into the future depends on the degree of freedom for unconstrained urban development provided within the region. In case, areas with a high suitable areas for conversion into built-up areas are not available in close proximity to for instance existing urban agglomerations, urban growth might leapfrog to dislocated areas. This is partially the case in Mumbai, where the existing peninsula is almost saturated by urban development. Urban development is only possible in the proximity of neighbouring cities across the Thane creek or further north.

Extrapolation combined with varying levels of constraints for urban development is reflected in the observed growth rates in the developed scenarios. The projected growth rates are highest in Dhaka during the interval 2015-2025. Here the urban footprint is expected to growth by more than 80%, which is somewhat confirmed by the current growth expectations in which the city's population is expected to almost double in that same period. On the other hand, the lowest rate is projected for Seoul in the interval 2045-2055: less than 5%. A more detailed analysis of the individual growth scenarios can be found in Appendix A3 as well as in Veerbeek et al (2014).

## 5.2.2 Assessing the urban flood extent and depth distribution

The urban flood extent describes the reach of a particular flood event in a built-up area, which is operationalized as the intersection of the flood extent and the urban footprint (Angel et al, 2007) at a particular year in the projected LULC change scenario. To summarize the different urban flood extents associated to the range of return periods into a single figure, the weighted mean urban flood extent $\bar{E}$ was calculated using the associated flood frequencies of a range of events:

$$(10) \quad \bar{E} = \frac{\sum_{i=1}^{N} \left( \frac{E_i}{T_i} \right)}{\sum_{i=1}^{N} \frac{1}{T_i}}$$

where $N$ represents the set of return periods for which the flood extent E has been calculated. Derived the GLOFRIS-dataset (see 5.2.1) the set $N=[10,25,50,100,250,500,1000]$. Using the mean annual urban flood extent does account for some loss of information, since disproportionate changes in the flood extent associated to individual events are absorbed into a single figure. Yet, comparison the urban flood extent for every single flood frequency or omitting a range of frequencies by comparing for instance only the extent associated to a 100 year event leads to either overcomplicating or oversimplifying the comparison. Nevertheless, the underlying data is available in Appendix A3. As an alternative, the weighted mean urban flood extent could have been calculated based on integration of the development of the urban flood extent over the range of feturn periods using a trapezoid form typically used to calculate the annual flood damages (e.g. Olsen et al, 2015).

To obtain a satisfactory trade-off between expressiveness and statistical significance in relation to the scale of the source material, flood depth distributions are binned by intervals of 0.5 meter and have been limited to 10, 100 and 1000 year events. Furthermore, the distributions have been calculated only for the current conditions (base year), the mid-term (2035) and the long term (2060). Computation of flood depth distributions associated to floods with different return periods and horizons is straightforward, but not been produced in order to keep the amount produced data manageable.

# 5.3 Validation

Although there are no comparable studies available to test the reliability of the produced outcomes, some procedures are available to at least provide more evidence of their robustness. To be more precise: to estimate how sensitive the produces flood extent and flood depth distributions are to changes in the size of the grid cells used in the individual datasets  the outcomes are based on: the GLOFRIS flood maps and the spatially explicit urban growth scenarios produced by the model of Veerbeek et al (2015). Depending on this sensitivity, the outcomes (e.g. the flood extent and depths) can be then be described as a range instead of a set of single values. If this range is limited compared to the actual values, the produced results can be considered as relatively robust. Additionally, better insight is provided in the consistency of the characteristics of the relationship between for instance the growth of a particular city and the development of the flood extent intersecting urbanized areas and the associated depth distribution.

The resolution of the GLORFRIS data set is defined by the resolution of the Global Watershed (GW) elevation data which is set at 30 Arc Seconds. This corresponds to grid cells of about 926m wide around the equator and somewhat lower for cities located further away from the equator (e.g. the cells size for Dhaka is about 832m). The urban growth scenarios on the other hand are based on Landsat TM and ETM data with 30m wide grid cells. Due to this substantial difference in resolution, intersecting the datasets to obtain flood prone urbanised areas will ultimately lead to imprecise outcomes. By matching the resolution of the two datasets and performing the assessment, alternative results can be produced. This can be done by either upscaling the LULC maps or downscaling the flood maps.

## 5.3.1 Initial setup

The initial setup consists of a superposition of the LULC maps produced from urban growth scenario and the flood extent derived from the GLOFRIS dataset. To dispose of noise and insignificant inundation levels, the minimum flood depth was set at 1 dm. The subsequent flood extent was then reprojected and resampled to match the corresponding 30m grid cells in the LULC maps. Finally, the urbanised flood extent was derived by intersecting all urban land cover with the flood extent. This was done for every year until the horizon of 2060. This extent was ultimately used in the calculations and comparison.

**Figure 18:** Built-up areas in the original 30m cell grid sized dataset (left) and resampled to 30 Arc Seconds (right).

## 5.3.2 Upscaling

To estimate the effect of aligning the LULC scenarios with the level of detail provided by the GLOFRIS flood maps, the built-up areas for the consecutive intervals between the base years and the 2060 horizon have been resampled from 30m to 30 Arc Second grid cells. Thus, the resolution of the resampled LULC maps matches those from the flood maps. Resampling was based on a majority rule, which means that depending on the frequency of occurrence, the resulting distribution of built-up areas might be changes. This is illustrated in Figure 18.

The method does not necessarily produce either an over- or underestimation of built-up areas since the produces results are dependent on the frequency and spatial distribution of the initial urban extent. Yet, depending on the city-size (i.e. the urban extent), resampling to a lower resolution might result in substantial deviations from the original urban form since these are more sensitive to the characteristics of local urban patterns. The outcomes for relatively small cities like Dhaka or Ho Chi Minh City might therefore differ significantly from those produced using the original 30m based LULC maps.

## 5.3.3 Downscaling

Apart from resampling the LULC maps to a larger cell size in order to match the resolution of the flood depth maps produced in GLOFRIS, a different approach is taken to downscale the flood depths maps to better accommodate the level of detail of the produced LULC scenarios. This was done by adjusting the flood depths using 1 Arc Second elevation data derived from the Shuttle Radar Topography Mission (SRTM) which corresponds to a cell size of about 30m at the equator (USGS, 2016). The procedure is rather straightforward:

1. Resample the 30 Arc Second-based GW elevation data and the GLOFRIS depth

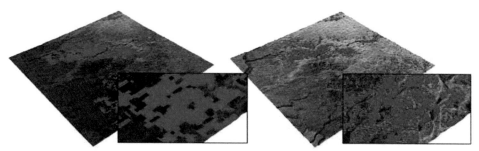

**Figure 19:** Flood extent superimposed on the GW elevation data (left) and the SRTM adjusted extent superimposed on the SRTM data (right)

     datasets to 1 Arc Seconds to match the resolution of the SRTM elevation data;

2. Calculate the difference in elevation between the GW and SRTM data;
3. Subtract the difference from the GLOFRIS flood depth maps and compensate for negative flood depths (i.e. set to zero);

In cases where flood depths are limited, this procedure can significantly change the flood contours and depths of the GLOFRIS datasets. This is illustrated in Figure 19 where the original flood extent of the GLOFRIS/GW elevation data differs substantially from the SRTM-altered extent.

An important advantage of this adjustment is that local differentiations in elevation levels are better expressed. Built-up areas located on relatively small ridges, mounts or other elevated areas are therefore no longer incorrectly marked as flooded, which might change the overall assessment of the exposure to flooding of a particular city. This might especially the case for floods with low return periods since the associated flood depths are usually small.

Yet, the adjustment might cause a possible bias towards a smaller flood extent since the extent is only contracted due to adjusted elevation differences. Expansion of the flood extent beyond the initial perimeter of the GLOFRIS has not been performed. One could therefore argue that the outcomes resulting from the 'downscaling' signify the lower boundary within the range the flood extent is estimated from the available data sources in this project. This is not the necessarily the case in relation resulting flood depths since adjustment using SRTM data works in both directions: depending on the differences in elevation between the datasets, the resulting flood depths can become smaller as well as larger.

## 5.3.4 Additional validation methods

In individual cases, it might be possible to compare the outcomes to those created using alternative data sources. For instance, for Dhaka alternative flood maps are available on a significantly higher level of detail than those produced by the GLOFRIS model. Yet, the hydrologic conditions these models are based on might differ from those used to produce the flood maps used in this study. This could results in a different flood extent and depth distribution while the estimated return period for the event might still coincide with an associated return period in the GLOFRIS dataset.

## 5.3.5 Validation results

### 5.3.5.1 Upscaling

The effect of the upscaling procedure compared to the initial setup differs substantially across the different case study cities. While for instance the average difference in urbanised flood extent for the different intervals between the base years and 2060 horizon is only 1.6% in Lagos, upscaling the urbanised areas in Seoul leads a flood extent that deviates more than 58% from the initial setup. Furthermore, the urbanised flood extent in for instance Lahore becomes substantially larger because of upscaling while in the Guangzhou-Shenzhen region it contracts. Yet, there seem to be a few general trends observable that can be explained quite easily:

First there is an issue of proportionality: for cities characterised by a small urban flood extent, scaling up the cells size can have a large effect. The resulting number of individual cells identifying flooded urban areas as a result of majority-based upscaling can change significantly. For instance when only small pockets of flooding exist, upscaling might omit these areas altogether. This example identifies a second determinant factor: the actual distribution of flooded cells in the initial setup. When for instance observing the effects of majority-based upscaling in cities like Istanbul, Seoul or Tehran where flooding occurs in the immediate vicinity of the rivers, the flood extent increases by 50%, 58% and 51% respectively when compared to the initial setup. This substantial difference might be related to the relatively large perimeter of the flood extent compared to more compact distributions where the flood extent is concentrated in a particular area. Due to this large perimeter, majority-based resampling of cells into urban or empty areas occurs more frequently and therefore has a relatively large effect on resulting number of upscaled flooded cells and the subsequent size of the urbanised flood extent. Cities with a more concentrated urbanised flood extent typically show smaller deviations from the initial setup. This can be perceived for instance in Jakarta, Mumbai and Ho Chi Minh City, where the average differences compared to the initial setup are 1.9%, 5.3% and -3.8% respectively.

Another trend that somewhat relates to the issue of proportionality is that for many cities the impact of upscaling diminishes for increasing return periods. This can simply be explained by the increasing flood extent that is in most cases associated to less frequent floods. This results in large contiguous flooded areas. In cities where these coincide with vast urbanised areas, the effect of upscaling diminishes since the areas were equally accounted for in the initial setup with a cells size of 30m.

Apart from these, almost no other trends can be identified and explained. For some cities, the impact of upscaling shows a large variability for the different growth iterations (e.g. Karachi) while for others, the impact remains almost constant (e.g. Mexico City). For some the impact decreases during increasing iterations (e.g. Shanghai) while for others upscaling has in increasing impact over time (e.g. Calcutta).

The final conclusion can therefore only be that the effects of normalizing the cell size of the flood maps and LULC maps by upscaling the majority-based upscaling the latter, has to be assessed per city individually. The characteristics of the patterns in which the pockets of built-up areas are distributed differ too much within and between cities to be able to make actual predictions about the effect of scaling up the cells on the urbanised flood extent. The evaluation of the effects of this method on the potential robustness of the outcomes has to be determined per city.

### 5.3.5.2 Downscaling

While for the upscaling procedure, the location of the inundated urban extent is the predominant factor determining consistent outcomes, for the downscaling procedure the flood depth seems to be the most prominent characteristic This can be observed for instance in Shanghai where for a flood with a 10Y return period, about 90% of flood depths are between 0.1-0.5m. For floods with a return period of 100Y and 1000Y, these percentages remain close to 90%. Even in the most extreme case: the expected inundated urban extent in 2060 during a 1000Y event, still 71% of that area faces these relatively shallow inundations, ranging between 0.1-0.5m. This makes the flood extent very sensitive to local spikes in the SRTM-based elevations that can easily exceed the levels reached by the inundation and led to a significantly smaller flood extent (see Figure 19). In the case of Shanghai, this lead on an average reduction of about 91% compared to the initial setup. As a result, the resulting range for the estimations of the flood extent in Shanghai compromise the significance of the outcomes to such a level that might need to be omitted.

Fortunately, Shanghai is an exception in set of cities used in the research. In many cities the impacts of downscaling on the flood extent are much less significant. Especially in for higher return periods, where flood depths are in many cases substantially higher, the reduction in flood extent is less than 10%. This can be perceived for instance in the Guangzhou-Shenzhen area, where the effects of downscaling for a 10Y flood event range between 21.3% for 1990 and 19.3% in 2060. For a 1000Y event though, this impact is reduced to 8.4% and 5.2% respectively. The trend of a declining impact for increasing return periods can be perceived for all cities. From all cities, the impacts of downscaling seem smallest for Dhaka. Here the reduction in flood extent is on average around 2.3%. This can be traced back to the relatively high estimated inundation depths; higher elevation found the SRTM data thus only have a very small effect on the calculated flood extent since a relatively large increase is required to approach the level reached by inundation.

The differences between Shanghai and Dhaka illustrate the main issue associated to the downscaling procedure: almost no general trends can be observed. The influence of the increasing urban extent due to urban growth has a mixed impact on the effects of downscaling. In Beijing for instance, the average effect of downscaling remains almost constant as a function of urban growth (around 9%), while in Jakarta the effect decreases until 2025 after which it increase again. The absence of general trends when assessing the impact of downscaling leads to the conclusion that the outcomes can only be asses per city individually.

### 5.3.5.3 Additional Data Sources

A rather obvious additional validation method would be to compare the produced out-

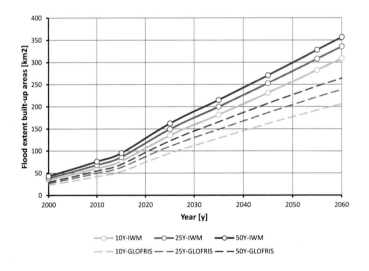

**Figure 20:** Mean, maximum and minimum flood extent as a function of urban growth

comes to those using flood data from alternative data sources. Yet, such a comparison is highly dependent on the availability of ample flood data for the different case study areas and for preferably a range of different return periods. Furthermore, the hydrologic conditions on which the flood maps are based on should preferably be identical. The availability of such data is limited. This is partly due to practicalities; due to the wide geographic extent in which the case studies are located, relations with many agencies have to be established to obtain those flood maps. If these are available, an additional question is if the data is actually. Due to the often sensitive nature (i.e. they might compromise real estate value and ground prices), they are often not disclosed. For the set of cities used in this study, the only available flood maps came for the city of Dhaka. Produced for another project, the Institute of Water Modelling (IWM) provided a set of 4 inundation maps for events with return periods of 1, 10, 25 and 50Y events, from which the 1Yl event was omitted since the lowest return period used in the project is 10Y. The level of detail of these flood maps was significantly higher than for the GLOFRIS dataset: 300m versus 832m (30 ArcSec) respectively.

After superposition of the IWM-produced flood map with the urban development scenario, the resulting mean urban flood extent for the range of development intervals was estimated 45%, 38% and 35% larger for the 10Y, 25Y and 50Y event respectively. While this seems substantial, this difference can be partly explained when comparing the two flood maps. Although the associated 3 return period are equal, the associated flood extent and depth distribution differ substantially. The total flood extent for the 10Y event for instance within the boundary tile of the Dhaka region is for instance about 97% larger than the estimated extent produced by the GLOFRIS model. While this factor drops to 71% and 56% for the maps associated to 25Y and 50Y events respectively, the flood extent is still considerable large. This means that the GLOFRIS-produced inundation maps cannot be simply considered as scaled up versions of the IWM maps which hampers their use for validation purposes.

Yet when observing the characteristics of the development of the urban flood extent as a function of the estimated urban growth, the behaviour is relatively similar. This can be observed in Figure 20, where apart from the overall lower levels, the shapes of the graphs are very similar to those produced by the GLOFRIS-based estimations. Apparently, the urban growth is the dominant factor which limits the effect of the differences in flood extent.

## 5.3.6 Conclusions from the validation

Although the down- and upscaling procedures affect the size of the urbanised flood extent and/or the depth distribution, the results are relatively robust. For most cities the deviations produced by these procedures are small compared to the influence of urban development on those features. This is illustrated in Figure 21, where the mean as well as the range of the estimated flood extent is shown as produced in the initial setup as well as the down- and upscaling procedure for flood events with three associated return periods. Furthermore, the ranges associated to the 10Y, 100Y and 1000Y do not overlap. Thus, the flood extent associated to frequent floods nowhere exceeds that of less frequent events, which would be inconsistent. In case of Guangzhou-Shenzhen, the ranges are become more clearly separable as a function of urban development which clearly indicates a much larger future urbanised flood extent for more extreme events. Yet, such behaviour is dependent on the local geographic conditions reflected in the GLOFRIS-produced flood maps and the urban growth scenarios.

Outcomes as depicted in Figure 21 are representative for the majority the investigated cities in this research. Exceptions are for instance Seoul (see Table 7) where especially the upscaling procedure affects the range of outcomes substantially while the impacts of downscaling are very limited.

This also occurs for the downscaling procedure as can be witnessed in for instance Shanghai (see Table 8), where there results deviate for more than 90% from the initial setup (compared to about 13% for the upscaled results).

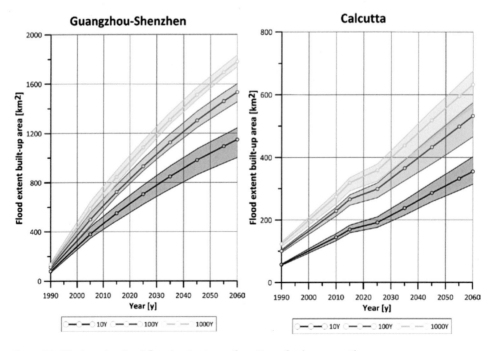

**Figure 21:** Dhaka urbanised flood extent as a function of urban growth

**Table 7:** Estimated flood extent for Seoul for the initial setup and alternative procedures

| Seoul | 1990 | 2005 | 2015 | 2025 | 2035 | 2045 | 2055 | 2060 | Avg. Dev. | Rank 2060 |
|---|---|---|---|---|---|---|---|---|---|---|
| GLOFRIS | 30.1 | 37.7 | 46.9 | 54.4 | 59.7 | 63.3 | 66.1 | 67.3 | | 13 |
| Upscaled | 45.6 | 67.5 | 72.9 | 81.6 | 90.2 | 99.1 | 105.5 | 109 | **58.20%** | 11 |
| Downscaled | 28.9 | 36.3 | 45.4 | 52.7 | 57.9 | 61.4 | 64.2 | 65.3 | 3.20% | 12 |
| Mean | 34.9 | 47.1 | 55.1 | 62.9 | 69.3 | 74.6 | 78.6 | 80.5 | | 13 |

Apart from, assessing the impacts of up- and downscaling compared to the initial set-up, the robustness of the relative ranking of cities based on the urbanised flood extent can also be examined. Out of 18 cities, the ranking for Seoul is relatively stable; it varies between rank 11 and 13 (see Table 7). The effect of downscaling in Shanghai on the other hand has a strong impact on the ranking: from rank 2 to 15. Especially Shanghai is an exception to the rule. The overall flood extent-based ranking of investigated cities remains relatively stable. In most cases the rank shifts no more than a single spot down or up the list.

An important aspect of the two validation methods is that they evaluate the outcomes in relation to the horizontal raster resolution (by upscaling) as well as the vertical resolution (downscaling).

With the available data it is impossible to estimate the likelihood which of the produced values better approximates the actual urbanised flood extent. As an alternative, the mean values have been have been calculated to act as the actual figure used in the overall assessment. Since for most cities the produced range is limited, this will not produce values that differ drastically from the initial setup but it does express some of the variability produced from the up- and downscaling procedures. The validation results for all cities can be found in Appendix A2.

## 5.4 Outcomes

The assessment focuses initially on the development of the urbanised flood extent,

**Table 8:** Estimated flood extent for Shanghai for the initial setup and alternative procedures

| Shanghai | 1990 | 2005 | 2015 | 2025 | 2035 | 2045 | 2055 | 2060 | Avg. Dev. | Rank 2060 |
|---|---|---|---|---|---|---|---|---|---|---|
| GLOFRIS | 97.3 | 291.3 | 335.8 | 399.1 | 444.5 | 480.1 | 511.7 | 525.9 | | 2 |
| Upscaled | 73.0 | 243.6 | 295.0 | 345.6 | 401.0 | 434.8 | 462.0 | 476.0 | 13.20% | 2 |
| Downscaled | 15.1 | 22.8 | 26.9 | 31.0 | 34.2 | 36.8 | 38.9 | 39.8 | **91.30%** | 15 |
| Mean | 61.8 | 185.9 | 219.2 | 258.5 | 293.3 | 317.2 | 337.5 | 347.3 | | 3 |

i.e. the patches of built-up area that are located within the projected flood extent. This analysis is performed by presenting the calculated extent as discrete outcomes, summarized in a ranking. Furthermore analysis has been performed by comparing the development of urban flood extent to the urban development projected outside the estimated flood contours. This provides evaluates the relative proportionality of the expected future flood exposure in relation to the projected urban growth. Finally, the analysis of flood exposure is complemented by an evaluation of the expected flood depth distributions, focussing especially on how the fraction of lower inundation depths in the investigated case study areas compares to higher depths.

## 5.4.1 Urban flood extent

When accessing the size of the urban flood extent, the Guangzhou-Shenzhen region clearly dominates all other cities (see Table 9). Currently, the urban flood extent covers about 668 km2 while in 2060 the area is expected to grow to 1417 km2, thus more than doubling in size. The extent clearly dwarfs the estimated extent of urban agglomerations like Shanghai or Calcutta which rank second currently and in 2060 respectively. Yet, in 1990 the estimated urban flood extent of the Guangzhou-Shenzhen area only accumulated to about 103 km2 , which would bring it much closer to agglomerations like Beijing or Mexico City for which the urban flood extent accumulated to about 93 km2 and 91 km2 respectively. The extensive growth of the urban flood extent reflects the extraordinary urbanisation rate of the Pearl River delta.

In the 2015 ranking, Guangzhou-Shenzhen is followed by Shanghai for which the cur-

Table 9: Top and bottom ranking based on urban flood extent for 2015 and 2060

| Rank | City | Flood extent 2015 [km²] | City | Flood extent 2060 [km²] |
|---|---|---|---|---|
| 1 | Guangzhou-Shen-zhen | 668.0 | Guangzhou-Shen-zhen | 1417.4 |
| 2 | Shanghai | 219.2 | Calcutta | 408.6 |
| 3 | Calcutta | 198.2 | Beijing | 341.4 |
| 4 | Beijing | 132.5 | Shanghai | 347.3 |
| 5 | Mexico City | 128.1 | Delhi | 288.6 |
| 14 | Mumbai | 23.0 | Mumbai | 56.1 |
| 15 | Lahore | 12.7 | Lahore | 51.7 |
| 16 | Karachi | 7.3 | Karachi | 19.0 |
| 17 | Tehran | 5.2 | Tehran | 11.3 |
| 18 | Istanbul | 1.0 | Tehran | 4.6 |

rent flood extent accounts for a little over 219 km2. Note that the results for Shanghai might be somewhat skewed due to the relatively low inundation depths. In this section on validation this is explained in more detail. Shanghai is expected to move to 3rd place in favour of Calcutta for which the urban flood extent is projected to almost doubles to 409 km2 in 2060. The position of Calcutta as one of the most flood prone areas is confirmed by the study of Nicholls (2008). Apart from Calcutta, India's capital New Delhi is also projected to face a significant increase in urban flood extent. Currently ranking in 6th place, the city's urban flood extent is projected to increase 2.5 times to about 289 km2.

One of the cities that rank surprisingly high is Beijing. Known in the past decades for its severe problems in relation to drought management and subsequent water supply problems, the city is currently ranked 4 for size of the urban flood extent, while it is projected to move to third place in 2060 almost tripling the urban flood extent to 341

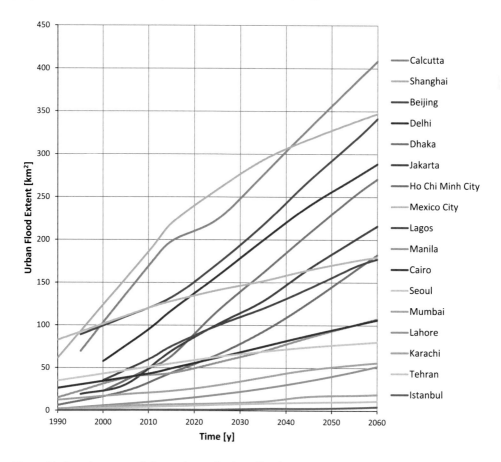

**Figure 22:** Development of the estimated urban flood extent over time as a function of urban growth

km2.

Among high ranked cities that are expected to be surpassed in the future is Mexico City, for which the estimated current urban flood extent of 128 km2 is projected to increase by 40% in 2060 which although substantial, is relatively modest compared to the projected increase of some of the other megacities. This moves the city from the current rank 5 to the 9th position in 2060.

At the bottom of the ranking, the rank order is stable for the projected interval till 2060. From all studied megacities, Mumbai, Lahore, Karachi, Tehran and Istanbul have the smallest urban flood extents. The relatively low position of Mumbai might seem surprising given the history of floods the city has experienced. Yet, these floods are almost exclusively due to monsoon driven local rainfall instead of overtopping of rivers and tributaries. In the case of Mumbai, the only flood produced by the GLOFRIS model was located in the Northeast region of greater Mumbai (see Appendix A4.15).

While the ranking order of cities remains relatively stable over time, two cities located in the middle portion of the list show more volatile behaviour. The urban flood extent of Dhaka is projected to increase more than 4-fold, from a current area of 64 km2 to 271 km2 in 2060. This moves the city from a current rank 9 to rank 6 in 2060. This shift is even more astonishing when comparing the size of the metropolitan area to some of the other megacities. The urban footprint of greater Dhaka currently covers around 218 km2. This is tiny compared to its closest ranked neighbour Jakarta (ranked 8) of which the urban footprint covers about 1292 km2, which is about 6 times Dhaka's size. A similar project growth of the urban flood extent is projected for Ho Chi Minh City, where the extent also increases about 4-fold to about 183 km2. In the ranking list the city is projected to move from a current rank 12 to 8 in 2060. By that time, about 24% of the Ho Chi Minh metropolitan region could be located in flood prone areas com-pared to 13% currently. For Dhaka, these figures might become even more alarming: while currently about 29% of the urban footprint is located in flood prone areas, in 2060 this figure could increase to more than 56%. Both Dhaka and Ho Chi Minh city are surrounded by rivers and vast wetland areas. Inundation is therefore expected beyond the direct vicinity of the rivers. The limited availability of land that is located outside the flood prone regions, limits the possibilities for urban expansion. Furthermore, the cities currently show very high density levels with hardly any places available for infill. Only when leapfrogging well beyond the flood prone regions, 'safe' urban growth is possible. Some of the lower ranked cities show similar magnitudes in the increase of the urban flood extent. Yet, for instance the outcomes for Istanbul are sensitive to scaling issues (see section 5.3 on validation) which might affect the projected 4-fold increase of the urban flood extent in 2060. The only exception might be Lahore. Yet the 4-fold increase might seem significant but only accounts for less than 9% of urban footprint in flood prone areas in 2060.

Apart from Mexico City, the lowest relative increase of the urban flood extent is projected in Seoul, where the area is estimated to increase by 46% in 2060. Surprisingly, the relative increase is also limited in Shanghai, where the urban flood extent is projected to grow by only 58%. Yet, the city is ranked 2nd throughout the complete projection and is expected to grow about 93% over that same interval.

The long term projections for 2060 and the subsequent differences compared to the current conditions require some additional insights on how the projected changes occur over time. Due to the differentiation in local conditions for each city, it is safe to assume the development of the urban flood extent does not progress in a linear fashion among all cities. This is illustrated in Figure 22, where the urban flood extent for all cities except the Guangzhou-Shenzhen area, is illustrated over the complete projection horizon. Since the size of the urban flood extent for Guangzhou-Shenzhen is of a completely different order of magnitude, were omitted from the figure.

From the figure, a differentiated yet characteristic behaviour can be observed which expressed in the shape of the different curves:

- Delayed increase for Calcutta. The urban flood extent for Calcutta shows a temporary departure from the strong, almost linear increase in the interval 2015-2025. Calcutta also shows the strongest growth from all the cities depicted in Figure 22.
- Exponential increase. Especially in Beijing, Manila but also for Lahore the urban flood extent tends to increase more rapidly as the years progress.
- Exponential decrease. Especially for Shanghai and Mexico City the growth of the urban flood extent seems 'flatten out' over time.
- Erratic development for Jakarta. Although the overall development is almost linear, the curve for Jakarta seems to oscillate over intervals of about 30 years.
- Bifurcation Beijing and Mexico City. During the interval between the base years, the urban flood extent of these two cities practically overlap. From 2012 on though, the observed trend for Mexico City is extended while the flood extent for Beijing shows a drastic increase.

What seems actually surprising though is the absence of strong deviations over time, expressed in much more irregular shapes of the curves in Figure 22. Apart from maybe Calcutta and Jakarta, the urban development in flood prone areas in many cities seems to progress in a rather trendwise fashion. Some of the irregularities in Figure 22 are somewhat hidden due to the extensive range of the graph required to depict the complete range between the urban flood extent of Istanbul to Calcutta. Yet, in many of the cases (e.g. in Lagos) a major portion of the river is located adjacent to an outward growing urban footprint. The urban flood exposure resulting from this development therefore grows therefore in a rather predictable manner.

Although the assessment primarily focuses on the potential consequences of urban growth, it is fairly easy to derive what portion of the projected urban flood extent is caused by existing urban areas. The contribution can be calculated by simply examining the size of the urban flood in the current year and calculating the fraction it occupies compared to the projected extent in 2060. Obviously, the contribution is dependent on the actual growth rate of the urban flood extent in the interval 2015-2060; the higher the growth rate, the lower the contribution. This becomes clear when examining Figure 22: Cities where the urban flood extent 2060 consists of more than half of the currently urbanised areas are Mexico City (71.2%), Seoul (68.4%) and Shanghai (63.1%). For all other cities, the contribution of the current flood prone urban areas is significantly lower. The lowest ranking cities are Dhaka (23.6%), Ho Chi Minh City (24.2%), Lahore (24.5%) and Jakarta for which the contribution is already significantly higher at 32.4%. For all other cities, the contribution is somewhere in the range of 35%-50%.

While the development of the estimated urban flood extent provides a good insight in the size of the expected problem, it fails to express how this development compares to the projected urban growth outside the flood contours. Yet, while the flood extent in a particular city might grow substantially, the increasing exposure might be relatively small compared to the overall growth of the city. Or, in other words, the trend-like development of that particular city might be directed to growth in relatively 'safe' areas: the city has a tendency for 'smart growth' without the need for readjustment. This characterisation can be measured by evaluating the growth of the urban flood extent against the growth outside the flood extent. This is simply operationalized as the ratio $\partial U_f \partial U$, where $\partial U_f$ represents the increase in urban flood extent over a period $t_i t_j$ and $\partial U$ represents the total urban growth over that same period. This growth characterisation might be a better performance indicator to express the relative changes in flood risk development over time. The metric is illustrated in Figure 23, where the two growth rates are set against each other for four consecutive intervals: 2015-2025, 2025-2035, 2035-2045 and 2045-2055. To illustrate the difference between a city where growth is primarily occurring outside the flood contours and a city where growth occurs within the flood contour, Shanghai and Lahore have been used as examples. The outcomes for Shanghai are located clearly above the isoline that illustrates

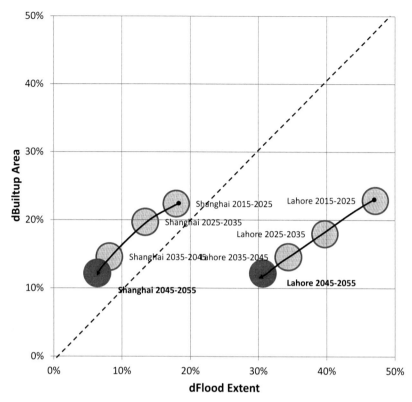

**Figure 23:** Growth rates of flood prone against flood secure urbanised areas for Lahore and Shanghai

an equal growth rate within and outside the flood contours, while Lahore is located clearly below the isoline.

When exploring these ratios for the interval between 2015 and 2060, the performance of a number of cities is comparatively poor, i.e. $\partial U_f \partial U$ shows values significantly larger than 1. This is especially the case for Dhaka where $\partial U_f \partial U$ equals 3.84. In other words, the projected growth rates within the flood contours are almost 4 times as high as outside the flood contours. Apart from Dhaka, cities with relatively high values for $\partial U_f \partial U$ are Ho Chi Minh (2.18), Lahore (2.08), Beijing (1.59) and Jakarta (1.58) perform relatively bad. Cities where urban growth rates outside the flood contours outperform those within (i.e. $\partial U_f \partial U < 1$ ) are for instance Shanghai (0.84), Seoul (0.85) and Mexico City (0.88). Finally, cities where the growth rates within and outside the flood contours are proportional ( $\partial U_f \partial U \rightarrow 1$) are Lagos (0.98) and the Guangzhou-Shenzhen area (1.06) . Other cities that show a tendency for proportion growth are New Delhi (1.13) and Cairo (1.15). Also for Calcutta the projected growth is relatively proportional ( $\partial U_f \partial U = 1.17$) which is somewhat unexpected since in absolute terms, the city's flood extent is rapidly expanding (Table 9).

Since $\partial U_f \partial U$ indicates the proportionality of two growth rates, the metric is calculated over a particular interval (e.g. 2015-2060). Yet, the projected growth rate of the urban footprint is not necessarily constant (as is the subsequent urban flood extent). When $\partial U_f \partial U$ is calculated over smaller intervals, both the components as well as the resulting ratio can change over time. This can be observed in Figure 23 for Shanghai as well as for Lahore. First of all, the growth rates become lower as time progresses. This seems intuitive since in order to keep the growth rates constant, the urban footprint needs to grow exponentially. In Shanghai as well as Lahore, the growth rates decline, albeit not at a constant rate: the biggest growth deceleration in Shanghai can be observed between the 2025-2035 and the 2035-2045 interval, while for Lahore this occurs between the intervals 2015-2025 and 2025-2035. In Figure 23 this is represented by the changes in distance between circles. More importantly though is the direction in which $\partial U_f \partial U$ is evolving over time. In Figure 23, the lines connecting the different intervals are nearly parallel to the isoline indicating a proportional development of urban flood extent. The slope of the line for Lahore is slightly lower though, indicating a move towards a more proportional development of flood extent. Furthermore, the line for Shanghai has the shave of a concave downward curve, indicating that the development trend of the ratio is changing over time. While for Shanghai and Lahore the

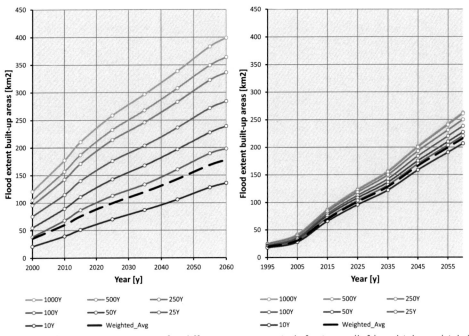

**Figure 24:** Flood extent over time for different return periods for Lagos (left) and Jakarta (right)

behaviour is rather gradual, for some cities more erratic behaviour can be observed (see Appendix A4).

### 5.4.1.1 Flood extent differentiation

As already explained in section 5.2.2, the assessment is focussing on the weighted mean urban flood extent. While for comparative purposes between different megacities this metric is effective, it dissolves potentially large differences between the different flood extents associated to the return periods it is composed of. For most of the case study cities in this research, the urban flood extent associated to a 10Y return period is substantially smaller than the extent associated to a 100Y or even 1000Y event. Yet, in cases where the floodplain is constrained by for instance relatively steep river banks, increasing discharge levels hardly affect the resulting flood extent. Both cases are illustrated in Figure 24, where the urban flood extent as a function of urban growth is illustrated for Lagos and Jakarta for 7 different return periods, ranging between 10Y and 1000Y. Furthermore, the weighted mean is included that is used throughout the analysis.

Figure 24 clearly illustrates the difference between the broad distribution for Lagos and the quite narrow distribution for Jakarta. Obviously this has some consequences for the robustness of the conclusions derived from the presented outcomes. For cities like Jakarta, the outcomes based on the weighted mean urban flood extent are more reliable since the outcomes are less sensitive to the occurrence of events with higher return periods.

Besides Jakarta, cities that fall within this class are: Cairo, Istanbul, Ho Chi Minh City, Shanghai and Tehran. Cities show relatively broad distributions are (besides Lagos): Karachi, Mexico City and Seoul. All other cities in the research show distributions that are moderately broad. In Appendix A3, the urban flood extent for the complete range is illustrated so depending on the application the flood extent can be included as a range instead of a single figure.

## 5.4.2 Flood depth distribution

Although the flood extent used for this research is a good indicator for acquiring knowledge about the potential scale of the flood problem in highly urbanised areas, some analysis about the flood depth distribution is required to provide additional information about the magnitude of the problem and the associated level of disruption cities might endure. Even though for the calculation of the flood extent a 10cm depth is taken threshold to omit very shallow flood depths, the flood magnitude might differ

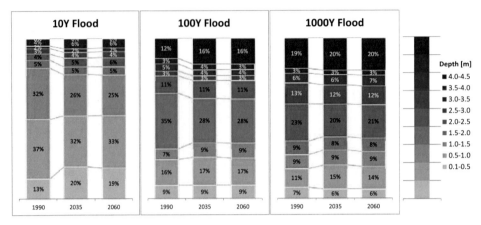

**Figure 25:** Estimated flood depth distribution for Guangzhou-Shenzhen for different years and return period.

substantially within and between the calculated urban flood extents.

To overcome this issue, the flood depth distribution has been calculated for the complete range of return periods and future increments until 2060. Yet, the amount of data this involves is not suitable for publication and will be available online as an additional set of tables. To provide some of the key data and subsequent interpretation, Appendix A3 includes the flood depth distributions for the 10Y, 100Y and 1000Y events for an initial base year as well as the medium term (2035) and long term horizon (2060). These are presented as as part of the data provided for every case study city. An example is provided in Figure 25 for the Guangzhou-Shenzhen region.

The assessment of the flood depth distribution focusses on two main issues:

1. How do the flood depth distributions compare between events with different return periods, e.g. 10Y, 100Y and 1000Y events?
2. How does the depth distribution develop for the applied urban growth scenarios?

When assessing the flood depth distribution for the metropolitan areas used in this study, a first observation is that for a number of cities the depths are relatively limited. This is especially case for Beijing and Shanghai. Especially in Beijing, more than 95% of the flooded urbanised areas face inundations of 50 cm or less. Only in the most extreme case: the 1000Y flood event in 2060, this level drops to about 93%. In Shanghai the distributions are also dominated by lower inundation depths albeit to to a smaller degree: Minor inundations below 50cm account for about 80 to 90% of the locations for frequent (10Y) to extreme (1000Y) events respectively. Application of the urban growth scenario drops the proportion of lower inundations with about 15% when measuring from the lower base year 1990. Yet, overall the effect of urban growth in Shanghai on a shift towards higher inundation depths is relatively moderate. Note that the observations of low inundation depths might lead to errors in the calculations of the urban flood extent. An extensive description about this issue is provided in the section 5.3 on validation.

Apart from metropolitan areas that are characterised by low inundation depths, the opposite is the casein for instance Mumbai. For a 10Y event, 62% of the urban flood extent is inundated above 2m. For the 100Y and 1000Y events, this percentage increased to 64% and 85% respectively. Yet, these latter figures are less alarming since such the exceedance probabilities are substantially lower. Interestingly, the applied urban growth scenarios do not shift these fractions significantly. In 2060, the portion of floods exceeding 2m is projected to be 69%, which marks only a 7% increase.

Like for Mumbai, urban growth seems to have a very limited effect on the flood depth distribution. This is especially the case for Delhi, Dhaka, Lagos and Mexico City. Yet, like unlike Mumbai inundation depths shift towards higher levels for events with higher return periods (i.e. 100Y and 1000Y events). In Cairo, Dhaka, Seoul and the Guangzhou-Shenzhen region this shift leads to significantly larger share of high inundation levels.

A typical example where both the impact of urban development as well as events with higher return periods cause a significantly bigger proportion of high inundation levels is Karachi. In many of the remaining cities, this relationship is less clear. In Calcutta for instance the distributions seem relatively unstable. Shifts in the flood depth distributions can be observed due to increasing return periods as well as urban development. Yet these shift do not show clear trends (e.g. linear or exponential increase). Similar shifts in distributions can for instance be observed for Jakarta, Lahore, Manila as well as for Ho Chi Minh City (see Appendix A4.7, Fig. 47) where urban development in medium term (2035) initially increases the proportion of lower flood depths (< 50 cm) after which the contribution of lower flood depths decreases again. To better evaluate how changes for especially this class of cities affect individual regions at some future point in time, inspection of the actual flood maps superimposed on the urban growth scenarios might be required. This requires the use of the actual data sources that are available on request.

Finally, when assessing the shifts in the flood depth distributions it important to realise that the figures indicate the fractions over the total flood extent. Only when combined, the actual size and magnitude of the potential future flood problem can be evaluated. Obviously, to actually assess the actual flood risk for an urbanised area, additional information is required including ample information about the receptor (i.e. the built-up areas) and the sensitivity of the receptor that determines the actual impact of a flood event. These issues are discussed in more detail in section 5.7, focussing on the interpretation of the outcomes.

## 5.5 Evaluation and Conclusions

The fact that the urban flood extent increases when cities grow is fairly obvious and well established as a main contributor to increasing flood risk in urbanising deltas. The essential question in this research though is to characterise the contribution of urban development to increasing flood risk; to assess if there are cases where urban development causes a disproportionate increase of flood risk. From the presented results, the cities where this seems especially the case are Dhaka, Ho Chi Minh City and Lahore. This is illustrated in Figure 26 for all megacities in this study for the interval 2015 till 2060, where these three cities are clearly outliers compared to the distribution of all other cities. The projected growth of the urban flood extent in Dhaka exceeds the pro-

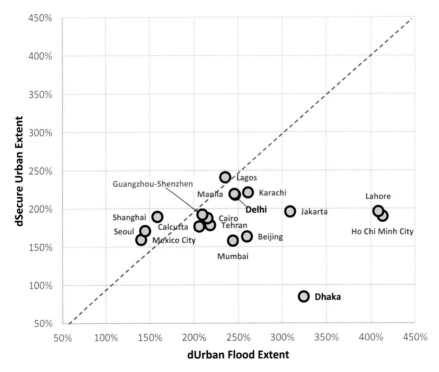

**Figure 26:** Comparison between estimated urban growth rates within and outside the average flood extent for the interval 2015-2060.

jected growth outside the flood contours by a factor of almost 3.8:1. For Ho Chi Minh City and Lahore this ratio is lower (2.2:1 and 2.1:1 respectively) but still substantially larger than for all other megacities.

Apart from the tendency to develop mostly in flood prone areas in relative terms, it is also the actual total share of the urban footprint that is particularly high in Dhaka and Ho Chi Minh City.

In 2060, almost 50% of the Dhaka metropolitan urban area is projected to be located in flood prone areas. For Ho Chi Minh City, this portion might be as high as 24%. The only other megacity in this study showing a similar level of exposure is Calcutta, where this portion in 2060 could be as high as 35%. Yet, Calcutta's growth is much less skewed towards growth in flood prone areas (see Figure 26), which leads to the conclusion that the high level of flood exposure is mostly a heritage from current conditions.

Interestingly enough there are no cities where the development is skewed to a similar degree in the other direction. Apparently, there are no cities where the vast majority of development is projected in 'safe havens'. Although the projected development in Mexico City, Seoul and Shanghai shows a clear trend in flood sensitive development, these ratios are nowhere close those found for 'flood prone'-development as for instance in Ho Chi Minh City. The maximum ratio between development outside and inside flood prone areas for the 2015-2060 interval is found in Shanghai, with a ratio of 1.2:1, which is approaching proportional development. Yet, due to the low flood depths in Shanghai the growth of the urban flood extent is possibly overestimated. The actual growth might be much more skewed towards flood sensitive development. This is also an issue in Beijing, for which the value might be less skewed towards a disproportionate development in flood prone areas.

When comparing the observations to the population size of the actual cities, there seems to be no direct relation between the population size of the metropolitan areas and their projected development in flood prone areas. Yet, Ho Chi Minh City and Lahore are the smallest megacities in this research with estimated populations of 10.1 and 10.3 million inhabitants respectively (Demographia, 2016). With 16.2 million inhabitants (ibid), Dhaka is significantly larger though although it still only ranks 11th in terms of population size when compared to the other megacities in this study. Apart from the absence of a relation between population size and urban flood extent, there also seems no relation between the population density and urban flood extent.

Here Dhaka dominates the statistics with 44.1 thousand inhabitants per km2 while Ho Chi Minh City and Lahore score significantly lower with 6.5 and 13.1 thousand inhabitants per km2 respectively. On the one hand, the absence of relation between population size and density and absolute or relative growth of the urban flood extent seems intuitive, since growth characteristics in relation to water bodies are not necessarily determined by these two properties. Yet, this again provides evidence for the absence of a bias in relation to possible errors due to the sheer area of the urban footprint or the level of 'compactness' of the urban footprint in relation to the urban flood extent. When decomposing the outcomes of Figure 26 into smaller intervals, the observed development trends for Ho Chi Minh City and Lahore (see Figure 23) can be almost linearly projected into the future; $\partial U\_f \partial U$ remain almost stable. In other words, no relative improvement of the conditions is expected in the future. This differs for instance from Dhaka, where the dispositional urban growth rate within and outside (i.e) seems to diminish over time, ie. $\partial U\_f \partial U$ becomes smaller. Potentially problematic trends can also be observed for Manilla, which seems to move from relative flood sensitive development towards flood prone urban development from 2025 on.

## 5.5.1 Ranking in relation to coastal flooding

Since the report by Nicholls et al (2008) in which future port cities are ranked in terms of exposure to sea level rise related impacts is cited extensively and had a significant impact in creating urgency, the development of global partnerships and programmes to increase future flood resiliency, it might be useful to compare the ranking produced in this study to that presented in the report. Apart from the already earlier discussed differences in approach and methodology, a major difference is that the research of Nicholls et al (ibid) is not limited to fast growing megacities; the ranking includes all coastal urban agglomerations with a population size of 1 million or more in 2005. This means that the assessment comprises of a substantially larger set of cities compared to the research presented here (136 vs 18 cities). Major urban agglomerations like Tokyo-Yokohama or New York-New Jersey are therefore included while they are omitted from the study presented here due to the fact that their growth rates are very low. Furthermore, Nicholls et al (ibid) developed two separate rankings based on asset value and population size respectively. The outcomes presented in the study here, are primarily based on the size of the urban flood extent.

Nevertheless when comparing the rankings, some similarities can be found. To illustrate these, the ranking of megacities that appear in both studies is presented in Table 10. Note that the ranking used from NIcholls (ibid) is the one based on exposed population size.

In both studies, cities like Calcutta, Dhaka Guangzhou, Ho Chi Minh City and Shanghai are high up in the rankings. This means that their already high exposure to coastal flooding is further aggravated by additional high exposure to riverine floods. Note that

**Table 10:** Comparison of the ranking of flood exposed cities

| Rank Nicholls' coastal. FR 2070 | City | Ranking Urban Flood Extent 2060 |
|:---:|:---:|:---:|
| 1 | Calcutta | 2 |
| 2 | Mumbai | 14 |
| 3 | Dhaka | 6 |
| 4 | Guangzhou-Shenzhen | 1 |
| 5 | Ho Chi Minh City | 8 |
| 6 | Shanghai | 4 |
| 15 | Lagos | 10 |
| 20 | Jakarta | 7 |
| 45 | Manila | 11 |
| 48 | Karachi | 16 |
| 59 | Incheon (Seoul) | 13 |
| 65 | Istanbul | 18 |

while both events can be treated as independent drivers of future flood exposure, the potential effects when both events coincide could be even more devastating. The ranking of Guangzhou on 4th place in the study by Nicholls et al (2008), is mainly due to the fact that they urban extent is limited to the administrative boundaries of Guangzhou. If adjacent agglomerations like Foshan, Dongguan, Zhongshan and Shenzhen would be included the metropolitan area would certainly rank first as well. In both studies, Dhaka is rapidly moving up the ranks; it ranks 14th in 2007 in the Nicholls et al-ranking (2008) compared to a 10th place in 2005 in this study. A similar jump in both ranks is observed for Ho Chi Minh City.

In the lower ranks in both studies we find cities like Istanbul, Seoul and Karachi. Mumbai is ranking high (2nd) in exposure to coastal flooding but relatively low in riverine flooding (14th). Yet, no cities ranked high for exposure to riverine flooding are ranked low for coastal flooding. In this respect, this study does not move cities significantly up the ranking. Yet, the study presented here does introduce a set of cities that might be currently be outside of the scope of climate change induced future flooding. Cities like Beijing or Delhi that might not be necessarily make it in the lists produced in mainstream media, might find their position readjusted by some of the outcomes of this study.

## 5.5.2 Consequences for urban flood risk management

Although the outcomes of this study clearly illustrate a case for urgent action to keep future urban flood risk in check, the presented outcomes do not necessarily provide a clear pathway for mitigating those risks. Sets of cities showing similar growth trends or characteristics do not necessarily require similar approaches in flood risk management since local conditions between cities differ dramatically. Thus, 'off-the-shelve' policy package would probably not suffice in keeping the projected future flood risks at acceptable levels. Instead tailored solutions are required that are adjusted to hydrological, spatial as well as the socioeconomic characteristics. Nevertheless, the outcomes of this study are not completely without consequences. The presented growth characterisation (see Figure 26) in combination with the projected size of the urban flood extent does provide insight in the requirement to alter the current spatial growth trends of some of the cities. For cities characterized by flood prone development (located below the isoline in Figure 26), a zoning policy where development is excluded from flood prone areas might be an important first step. In turn, cities characterized by flood sensitive development (located above the isoline in Figure 26) might be better served by flood mitigation measures that limit the flood hazard and subsequent flood extent. This analysis seems so to provide an initial lead or direction for future flood risk management policies.

The outcomes also provide some perspective for cities that currently already suffer from flood impacts. Especially the megacities located in India (Calcutta, Mumbai and New Delhi), Bangladesh (Dhaka) and Pakistan (Lahore, Karachi) have been coping with frequent and extensive riverine floods. Especially those cities in which the trends point at an increasing flood prone development might need to recognize the urgency to control their future growth.

In other cities though, even though the future urban flood extent might grow substantially, the rapid development outside the floodplains implies a relative improvement of the situation. Among others, the development of the regional GDP might far exceed the expected growth of future flood damages and might thus provide some means of coping and recovering from flood events that might currently prove to be catastrophic. Although in practise flood compensation does not necessarily flow from public or private resources in flood free areas to those affected by floods, it at least puts the development of future urban flood risk into perspective (Veerbeek et al, 2014). Nevertheless, flood damages and the resources required for recovery are in most cases not distributed equally over the urban population. In many cases, uncontrolled urban growth in floodplains often concerns the poorest portion of the population. Taking control over the unstoppable growth of shantytowns and slums therefore should be high on the agenda of many rapidly growing megacities in the developing world.

## 5.5.3 Responding to increasing urban flood risk

The combination urban growth induced flood risk exacerbated by climate change presents urban planners and decision makers with challenges they never faced before. Moreover, the existing urban and planning theories seem to be unsuited to provide guidance to understand these trends (e.g. Blanco et al, 2009) and to develop strategies to respond to them. The variation in the evolution of urban flood exposure between the different cities suggests that pro-active management of urban growth is in many cases likely to dampen the projected flood risk.

Yet, in the developing world the application of policies and associated instruments to direct or constrain urban development are often failing. Zoning regulations are often not enforced due to capacity limitations of municipal institutions, neglect, favouritism and corruption. Furthermore, many cities actively simulate investments, often at the expense of delineating strict development constraints to ensure a sustainable growth of the city. The pressure to maintain short term growth rates and associated GDP development often outweighs with long-term goals and objectives. Other issues include lack of proper instruments, legislation, knowledge as well as outdated cadastral maps. But even in cities where these issues less prominent, the urgency for enforcing a flood sensitive development policy is often lacking due to the fact that strategic flood risk management is targeted at limiting future flood losses. Since future impacts are often perceived as somewhat abstract or at least assessed as highly uncertain, little action is taken. This is somewhat ironic since a substantial portion of the cities in this research have experienced catastrophic flood events in the past (e.g. Dhaka, Lahore, Mumbai and many others.). What seems an important obstacle for cities to take action is a clear understanding of the impact of strategies and measures over an extensive period of time; no reference case is available from which the benefits of long term flood management strategies can be estimated. At the same time though, future costs required to repair "past mistakes" like the proliferation of urban development in flood prone areas, are typically ignored. Regrettably, such actions are often only taken after a disaster occurred when sufficient societal and political momentum exists.

Obviously, a pro-active flood sensitive urban development strategy is merely a component in a much larger ensemble of possible actions to increase the flood resiliency of metropolitan areas. In addition or as an alternative to flood sensitive planning the following measures should be considered in mitigating future fluvial flood risks in these megacities:

- Capital investments in flood prevention (e.g. dams, river embankments and river widening). An unprecedented shift in long-term investments (mainly in the public domain) in these megacities will be required using large scale, engineering solutions Compared to most Western cities, fast developing cities need to develop new flood defence infrastructure and not simply to preserve or adapt the old, existing systems;

- Changes in norms and regulations to facilitate autonomous actions (e.g. altered building codes, technical standards);
- Economic instruments (such as transfer development rights) directly and indirectly provide incentives for anticipating and reducing impacts;
- Infrastructural planning incentives; the availability of basic infrastructure such as roads, drainage, and electricity networks are key drivers for urbanization and thus provide opportunities to indirectly steer urbanization;
- Changes in individual behaviour (private with possible public incentives);
- Emergency response procedures and crisis management (mainly public);
- Risk sharing and transfer mechanisms (insurance), loans, public-private finance partnerships.

Clearly, the actual choice of instruments depends on a comprehensive decision framework in which a set of multi-criteria needs to be met. Furthermore, in most cases flood risk covers only fraction of a wider climate adaptation and mitigation strategy.

## 5.6 Extending the outcomes: CC-sensitivity

In this chapter an attempt has been made to assess the impact of urban growth on future flood exposure. While the impact in many cases seems disproportionately large, the question remains how the projected changes compare to future flood risk amplified by climate change. This becomes even more appropriate due to the long term horizon used for the urban growth projections in this study, which is well beyond the customary 20 year timeframe used for most long term urban planning policies (e.g. Gaubatz, 1999; Levy 2015; RAJUK, 1997). This aligns the projections with climate change projections which typically use 2050 as medium term and 2100 as a long term horizon (e.g. Nakicenovic et al, 2000). Nevertheless, the integration of climate change-driven future scenarios, the subsequent changes in rainfall patterns as well other changes in the hydrological cycle would have fare reaching implications for the scope of this study. This option was therefore dismissed.

In regards to the flood extent produced by riverine flooding, there might be an alternative which could at least provide a crude approximation of how the impacts of urban growth and climate change-induced amplified river discharges compare. A typical aspect of estimating climate change induced riverine flooding is the shift of flood frequencies. Depending on the climate change scenario, return periods associated to

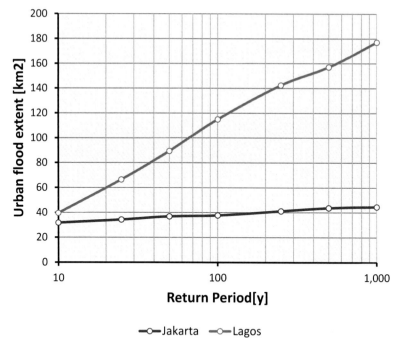

—○—Jakarta  —○—Lagos

**Figure 27:** Growth of the urban flood extent for 2010 over increasing return periods

specific river stages often become smaller (e.g. Hirabayashi et al, 2008).  For instance a river stage and associated flood extent that currently occurs once in a hundred years, might in 2100 occur once in 10 years.

Depending on the actual difference between river stages associated to a set of return periods  as well as the terrain characteristics of the adjacent floodplains, the subsequent flood extent might in some cases be only marginally different. In other cases though, the difference might cover hundreds of square kilometres over a particular river stretch. If that section is extensively urbanised, the urban flood extent will mostly likely differ comparably. As already explained in section 5.4.1.1 on flood extent differentiation, the flood maps used range between events associated to 10Y and 1000Y return periods. To resulting changes in the urban flood extent are illustrated for two extreme cases in Figure 27, where the estimated urban flood extent for 2010 is shown for all available return periods from the GLOFRIS derived dataset. The first case is Jakarta, where the flood extent hardly changes for increasing return periods. The urban flood extent associated to a 1000Y event is only 50% larger than the extent associated to a 10Y event. Although this does not express anything about the associated flood depths or the subsequent impacts, it does imply that the characteristics of infrequent river floods in Jakarta do not differ dramatically from more frequent events. In the case of Lagos, which represents the opposite end, the difference is close to 350% which is obviously a considerable difference. As a consequence, one might conclude that Jakarta is less sensitive to shifting flood frequencies than Lagos.

If climate change results in shifting flood frequencies, the current differences in flood extent associated to events with different return periods might act as a proxy indicator for how sensitive a particular city is to those shifts. Obviously, the indicator has many flaws. For instance, there is no indication how much the return periods would actually shift. Apart from the differences in changing rainfall patterns fro the different meteorological regions the river basins are located in, the river basin characteristics including the changing conditions for contributing tributaries differ between the cases used in this study. Despite these shortcomings though, the level of dispersion found in the estimated urban flood extents does provide a crude initial indicator.

To compare the range in the flood extent to the projected changes due to urban growth, the estimated increase of the weighted mean urban flood extent can be used (section 5.2.2, eq 10) for the interval 2010-2060. In Lagos for example the increase in the weighted mean urban flood extent in the 2010-2060 interval is about 117.7km2. Yet, the estimated difference between the urban flood extent associated to the 1000Y and 10Y events is about 137.7 km2. Hence, the sensitivity to shifting flood frequencies is relatively high compared to the increase due to projected urban growth. In Jakarta on the other hand, the urban growth affected increase for the 2010-2060 interval is about 185.6km2 which is considerably larger than the range of the 2010 flood extent, which is estimated at 13.8 km2 between the 1000Y and 10Y events.

Consequently, by means of introducing a simple equation that expresses the ratio between differences between the growth and flood frequency associated urban flood extent, the growth ratio $GR$ is conceptualised as:

$$(11) \quad Gr \quad \frac{\overline{E}_{t\,max} \quad \overline{E}_{t\,min}}{E_{N\,max} \quad E_{N\,min}}$$

where $\overline{E}$ corresponds to the weighted mean urban flood extent associated $t$max: the maximum projection year 2060 and $t$min: the base year 2010. $E$ represents the urban flood extent for the associated maximum and minimum return periods in the range adopted from GLOFRIS dataset, represented as $N$max and $N$min respectively. The set that represents the projection years is $t=[2015,2020,...,2060]$ (see 3.5). $N$max corresponds to a 1000 year return period and $N$min to a 10 year return period from the set $N=[10,25,50,100,250,500,1000]$ (see 5.2.1). For values larger than 1, urban growth is dominating factor while for values smaller than 1 the city might be more sensitive to

**Table 11:** Ranking based on the estimated growth ratio

| Rank | City | GR | Rank | City | GR |
|------|------|-----|------|------|-----|
| 1 | Istanbul | 36.90 | 10 | Guangzhou-Shenzhen | 4.04 |
| 2 | Ho Chi Minh City | 14.10 | 11 | Delhi | 2.40 |
| 3 | Jakarta | 13.47 | 12 | Calcutta | 1.83 |
| 4 | Dhaka | 7.99 | 13 | Mumbai | 1.77 |
| 5 | Manila | 7.23 | 14 | Karachi | 0.93 |
| 6 | Shanghai | 5.85 | 15 | Cairo | 0.91 |
| 7 | Beijing | 5.41 | 16 | Lagos | 0.85 |
| 8 | Tehran | 4.87 | 17 | Seoul | 0.81 |
| 9 | Lahore | 4.74 | 18 | Mexico City | 0.67 |

shifting flood frequencies. Due to its composition, that the outcomes of Equation 11 are sensitive to the ranges used in the numerator and denominator in the fraction. For a given set of flood frequency-extent relations, the $GR$ would change when for instance the upper range of the return period associated flood extent is limited to a 100Y event instead of a 1000Y. Likewise, if the growth projection only covers 20 years instead of 50, the increase of the mean urban flood extent would also be limited. This means that outcomes can only be compared if the range of return periods and the growth projection interval are equal. In this study with its focus on consistent and uniform data sources, this is fortunately the case. Calculating $GR$ would therefore provide a crude estimation of the order of magnitude of the growth vs flood frequency-sensitivity for the different megacities. From this estimation a rank list could be derived to provide some idea how the resulting ratio compare.

The base data represented as a collection of graphs that show the urban flood extent for the range of return periods as well as for the projected urban growth, can be found in Appendix A4. This base data is used to calculate the growth ratios for each city. The outcomes are shown in Table 11.

For the top ranked cities in Table 11, urban growth drives future flood exposure while cities in the lower region are more sensitive to shifting flood frequencies. The outcomes are in many cases different from earlier presented stats (e.g. proportionality or ranking of urban flood extent presented in section 5.4.1 and Figure 26). While some cities at top of the ranking sustain earlier conclusions about a disproportionate sensitivity to urban growth (Ho Chi Minh City, Dhaka and to a lesser extent Jakarta), also cities appear high in the ranking that show more proportional urban growth. This is the case for Manila and Shanghai where the GR values are high, rendering them more sensitive to urban growth than to shifts in flood frequencies. Yet, their growth is nearly proportional within and outside the estimated flood extent (see Figure 26). The extreme GR for top-ranked Istanbul (36.90) is simply due to the absence of a major river and is only included for completeness. At the bottom of the list, the earlier observed trends are somewhat sustained. While the projected urban flood risk analysis already showed proportional urban growth for Lagos, Seoul and Mexico City, the growth seems relatively low compared to the covered range between frequent and infrequent events. For all these cities, the GR is significantly lower than 1.0.

Overall though, the outcomes show that for 13 out of 18 cities urban growth seems the dominating factor. Given that climate change-induced shifts in flood frequencies will be less extreme than the assumed shift across the full range between 10Y and 1000Y associated events, the dominance of urban growth as the contributing factor to the future flood exposure of cities is even bigger. Although ultimately, the projected growth rates might be lower than assumed, this dominance will only change if drastic trend changes in urbanisation rates are achieved. Up till now, there are no signs that this is occurring.

## 5.7 Discussion

One of the most obvious questions of a comparative study is if the outcomes are indeed comparable. While in terms of methodology, model setup, execution and assessment a uniform approach has been used the final issue is of course how to interpret the resulting outcomes regarding the urban flood extent, depth distribution and the subsequent relative change in relation to urban growth scenarios. A common mistake that might be derived from this research is to interpret the outcomes as a comparison of flood risk. Yet to cover the complete concept of risk, extensive information is required about the sensitivity of the different urban areas to floods, i.e. what are the direct and indirect consequences of a given flood to an urban area represented in the LULC maps used in this study. Yet, both within and between cities, the built-up areas differ significantly. This affects for instance the relation between flood depths and the subsequent damages which are expressed in so-called depth-damage curve. Apart from the extensive inventory that is required to obtain the data to estimate such impacts, a set of additional factors complicate this task. In case of the model used in this study, building typologies, household composition, services and numerous other factors that define the characteristics of urban units are ultimately summarized in 30m grid cells. Representing a wide range of urban characteristics at that scale level inevitably leads to significant level of schematization. Nevertheless, attempt have been made to estimate direct flood damages at this scale level (Chen et al, 2016) although these still required detailed. For instance, since housing is dominated by high rise buildings, impacts of a flood in Beijing are mostly related to indirect damages; inundation of low lying major streets and tunnels results in massive traffic interruption. Yet, cities in India face very different impacts. There, especially the vulnerable poorer regions of for instance Mumbai or Dhaka suffer direct damages to shanty towns, affecting the often improvised constructions. Yet, especially in Mumbai many of the shanty towns are located almost directly next to very affluent gated communities. This further complicates effective estimation of flood damages.

These considerations also lead up to the decision to treat the flood extent and flood depth distribution as two separate components. Integration of both factors (e.g. through some multiplication) into a single metric could result in a more integrated approach where the hazard would be described into a single figure. Yet, this could implicitly lead to assumptions about the level of severity of floods at a particular location and in turn the sensitivity of that area. In the end the decision was made not to feed biased interpretations and present the outcomes in a relatively raw form.

An important omission in the model is the explicit treatment of urban redevelopment, which in some cases can be considered as the largest contributor to the urban flood extent. In Mexico City, Seoul or Shanghai for instance, knowing which areas reach the end of their lifespan and are up for redevelopment or replacement might provide further insight into the size future urban flood extent. Integration of this aspect into the growth model requires a substantial extension of the dataset including the construction year of built-up areas (Veerbeek et al, 2010). Yet, in many cities urban redevelopment contributes a substantial portion of the building activity and in some cases might outweigh expansion into new areas. In China for instance, massive redevelopment of large residential neighbourhoods is taking place. These former areas dominated by Hutong or socialist housing from the 1960s are rapidly replaced by high rise offices and apartments with smaller footprints which might change the urban flood exposure.

To further expand the application of the outcomes it might be important to focus on a lower scale level and increase the expressiveness of the components used in the assessment: for instance to apply the results for actual flood impact estimations a lower scale level is required in which building types (e.g. high rise), different types of land use (offices, residential, etc.) and a proper distinction between buildings and infrastructure is made. Already to assess the number of affected people by the different flood scenarios, the density distribution of the urban population distribution has to be explicitly included in the growth scenarios. Yet, before extending the urban growth model and resulting spatially explicit growth scenarios, it is essential to first adjust the scale of the flood maps (30 Arcsecs) to those of the produced LULC maps (30m). Only then, consistent outcomes can be derived from the interaction between flood hazard and urban receptor.

Ultimately, the outcomes of this study need to find their way to the cities in question. A logical next step would be to assess impact of interventions and use of the model as explorative tool to identify development pathways to managing future flood risk also taking into account the institutional capacities, policies and economics. Within the CORFU project, some experience has been gained in the presentation, adoption and use of explicit urban development scenarios for the cities of Beijing, Dhaka and Mumbai. An important factor in a successful adoption of the developed material is to open the project to a contribution from relevant local institutions; e.g. city corporations, drainage authorities, water boards, etc. Instead of presenting the work as a 'fait accompli', local knowledge and data sources can be used to validate and enhance the urban flood extent and depth distribution for the range of events. Especially since the GLOFRIS data does not (yet) provide a level-of-detail that is expressive enough to fit the requirements of land use maps of individual cities, refinement is required. Within the CORFU project these activities have been performed for Dhaka (Khan et al, 2015). An important added value of the outcomes is the use in 'scenario thinking', which is in most cities hardly common practise. The growth scenarios and resulting urban flood extent are thus merely starting points of a more in-depth discussion of a flood risk informed future urban development strategy. An alternative to addressing cities individually might be to integrate the method and results in some of the global collaborative urban platforms like the International Council for Local Environmental Initiatives (ICLEI) - Local Governments for Sustainability, The 100 Resilient Cities Network from the Rockefeller Foundation, The C40 Cities Climate Leadership Group or other large urban networks were climate adaptation, disaster management or flood resiliency is a major topic. The outcomes might help these networks to move beyond the issue of merely creating an increased urgency for the topic of urban flood risk. The outcomes might be used for a common scenario-based strategy development, where flood sensitive urban growth becomes common practise.

# 6. Assessing the effects of urban growth on urban drainage

## 6.1 Introduction

In most cases, urban growth has a significant impact on the exposure and sensitivity of people and assets to floods. Urbanizing floodplains as well as building in close proximity of local depressions obviously increases the potential number of inundated properties. Especially when such areas are populated by the urban poor, subsequent impacts of inundations increase since properties lack sufficient protection to limit the consequences of floods. Yet, cities act not only as passive receptors but exacerbate the frequency and intensity of floods. Urbanisation and the associated increase of the urban heat island can cause changes in the microclimate and local rainfall pattern (e.g. Pathirana et al, 2014). More importantly though, extensive soil sealing as a consequence of urban growth decreases the infiltration capacity of territory and subsequently increases the volume of surface runoff resulting from local rainfall events (e.g. Shepherd, 2006). Typically, this leads to the flooding of roads and underpasses (Zhou et al, 2013) but in more severe cases to the inundation of properties including retail and industry. Such floods are especially a problem in urban areas where the pipe drainage system is underdeveloped in relation to the required capacity (i.e. the design rainfall event the system is developed for). This is often the case in rapidly developing cities in the developing world where an increasing disparity occurs between privately driven urban development and public spending on utilities. Often the investments in the pipe drainage network are falling behind in relation to the densification and subsequent increase of impervious surfaces and required drainage capacity from the network. Poor solid waste management, which causes blockages of inlets and outlets as well as substandard maintenance of damaged or broken components in the pipe drainage network only increase the consequences of this disparity. Finally, many of the megacities in this research project are located in a monsoon driven climate. During the wet season this leads to torrential rainfall events, where peak levels well exceed the drainage capacity required to limit excessive runoff and subsequent flooding of urban components.

If capacity in urban areas to infiltrate stormwater into the soil is limited, local storage of excess rainwater is essential to limit widespread impacts during peak rainfall events. Historically, many larger cities were composed of a patchwork of alternating built-up areas and green zones, often combined with surface water bodies. Furthermore, many cities originate from a set of expanding villages or small towns that merge together over time. This can be well observed in the Guangzhou-Shenzhen area which can currently be considered as an extensive urban agglomeration but in the early 1990s still consisted of a region of mainly rural areas in which medium-sized cities dominated their respective regions. Over time, many of the open areas have disappeared due to an increasing development pressure that drove up land prices. The rapid disappearance of surface water bodies can be well perceived in Dhaka. This process is illustrated in Figure 28, where eastward urban development also is creating an increasing pressure to fill-up the ponds, in Bangladesh referred to as *beels* (Alam et al, 2016). The figure shows an area east of the Badda neighbourhood in the years 2001, 2008 and 2015. While in 2001 the area still contains a large portion surface water, by 2015 most of the *beels* have disappeared. Many see the process of infill and subsequent disappearance of beels as the main cause of Dhaka's persistent urban flooding which seems to be getting worse every year (Bari et al, 2001; Iawid, 2004).

Yet, the actual drainage performance obviously differs per city and is dependent on

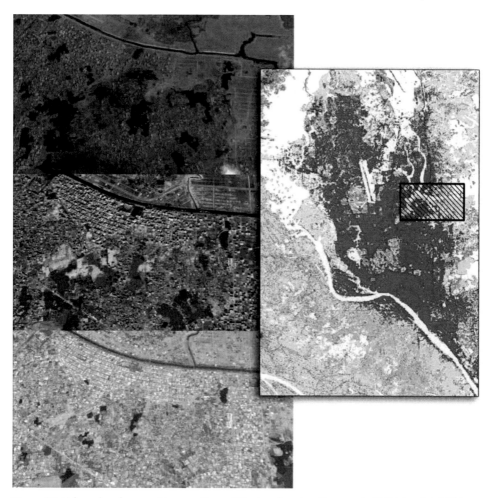

**Figure 28:** Urban development in a section of Dhaka, showing the area in 2001 (top), 2008 (centre) and 2015 (bottom). Photo's courtesy of Google Earth™

a multitude of factors including the distribution of patches of built-up areas. In some cases, the urban development might not significantly change the drainage character-istics. If level of fragmentation of alternating built-up and open areas remains stable in the future, the relative drainage performance, albeit over a larger area, might not necessarily alter. Future infill and urban expansion that eradicated the proportion of pervious areas thus cause a relative decrease of infiltration capacity; the overall drain-age characteristics have changed and urge the city to increasingly rely on constructed pipe drainage system.

Insight in the future drainage characteristics is essential in order to take appropriate ac-tions including proper land use management to ensure adequate urban drainage per-formance. Furthermore, for most cities no baseline scenarios for future urban drain-age performance have been developed in which urban growth is explicitly taken into account. Estimations of the impact of future strategies and interventions are therefore fragile. The application of regular 1d-2d coupled hydraulic models requires extensive data, calibration and computational resources for extensive areas like megacities. Ap-plications are therefore often limited to urban drainage units at neighbourhood level. Recently, some attempts have been made to cover vast urban areas using a simplified approach (e.g. Hénonin et al, 2013) but these still require datasets that exceed the typical output produced by spatially explicit LULC models including those use with-in this research. Alternatively, the future distributions of impervious surfaces can be estimated from the produced LULC maps after which key statistics and trends can be extracted to develop an appraisal of future drainage performance. Such an appraisal moves beyond the comparison of aggregate runoff coefficients that only provide infor-mation at a very rudimentary level. Since all case study areas are located on relatively flat terrain that extends well beyond the urban boundaries, the hydrologic system is dominated by manmade features; the identification of streams resulting from terrain morphology might therefore not change the outcomes substantially. This notion also affects the extent of the analysis on which the descriptive metrics are performed; us-ing river catchment or subcatchment might overemphasise the urbanisation of rural areas far away from the actual cities and therefore move the focus of the analysis away from the actual cities. Yet, the map extent used for LULC scenarios might work when analysing a single city but creates a bias in a comparative context since the borders are defined in an arbitrary way. Thus calculations for case study cities where the extent is rather tight might result in a significantly larger aggregate impervious surface ratio (ISR) than cities where the extent covers a wide buffer of rural areas around the met-ropolitan area. A alternative that provided consistent alternative was to use the urban footprint (Angel et al, 2005) as the analysis extent. Obviously, the urban footprint increases in size together with the growth of the urban areas in the projects. It therefore has to be calculated for all intervals the analysis is performed.

The following sections explore the development of impervious surface distributions for the previously described set of 18 rapidly developing megacities, using the same business as usual-based development scenarios as used for the assessment of fluvial flooding. Apart from a comparison, a more in-depth focus is provided for the city of Beijing which is witnessing unprecedented urban floods. Many of the trends in Beijing are illustrative for developments in other rapidly growing cities in this study. Furthermore, the coverage explains the metrics and results used in the analysis, which in aggregated form are provided in section 6.3 summarizing the main outcomes. The individual data including the maps with the projected impervious surface distributions as well as descriptive graphs are provided in Appendix B.

## 6.2 Operationalising future drainage performance through ISR

The LULC distribution and subsequent post processing to obtain a more expressive characterisation of the urban footprint provides a basis to estimate the impervious surface ratio (ISR) of each LULC class.

The resulting set of LULC class-ISR relations is a relative crude approximation of the actual ISRs existing in the actual case study cities. Application of sub-pixel regression methodologies (e.g. Yang and Lieu, 2005) could be applied to obtain the ISR ranges by using high resolution satellite imagery or other remote sensing data. Since this study involves future growth scenarios, such data is not available. Furthermore, extrapolation of ISRs based on observed data in for instance the base maps used for training the urban growth model introduces another class of uncertainty, since it introduces the question about the future character of the urbanisation (e.g. replacement of low-rise with high-rise buildings) which might alter the ISRs associated to specific classes of built-up areas. Yet, even without explicitly estimating the urban landscape class – ISR association for each individual city, the urban landscape classes can be ranked from high to low ISRs based on typical values associated to the estimated ground cover. This classification follows the often used SCS method (Soil Conservation Service, 1956; 1964; 1971; 1972; 1985) for land surface infiltration. The ISRs range between 0.05 for rural and 0.95 for highly urbanised LULC classes. For this research the following values are used:

- Urban built-up: 0.95
- Suburban built-up: 0.75

- Rural built-up: 0.6
- Urbanized open land: 0.4
- Captured open land: 0.2
- Rural open land: 0.1

Although this method is recognized to have a number of weaknesses, it is considered sufficiently accurate for the approximations required for this study. Furthermore, the results need to be comparable between different cities. Any method that increasing the expressiveness or accuracy of the estimations should therefore be appropriate and applicable for all cities. This adds an additional constraint to the use of additional data sources used in more elaborate estimation methods.

Urbanized and captured open land are LULC classes that represent patches of GI in the cities, that can contain areas for (temporary) water storage to relieve the city of some of the pressure on the drainage system during extreme rainfall. Yet, a city containing a large portion of urbanized and open land do not necessarily contain an evenly distributed GI; in theory all of the GI might be concentrated in a single green patch (e.g. central park in New York). Apart from inspecting the actual maps depicting the produced ISR distributions, insight in how contiguity of built-up areas can be provided by calculating the fractal dimension (FD). This metric expresses the level of fragmentation of patches of built-up areas as a scalar in the range [1,2] . The FD measures the log-log relationship between perimeter and area covered by all urban patches in the total analysis extent (Turner, 1990; Herold et al, 2005), i.e. the tile used for the base maps for the development of the growth scenarios (see section 3.3).

## 6.2.1 Case Beijing: extensive soil sealing due to concentric urban development

Although Beijing is mainly suffering from water scarcity (e.g. Zhang et al, 2011), the megacity also suffers significant impacts from urban flooding. For instance, in 2004 flooding led to inundation of flyovers in excess of 2 meters (Pan et al, 2012). In June 2011, unprecedented downpour lead to the closure of 3 subway lines due to flooded underground stations, cancellation of 76 bus routes and 40% of aircraft traffic. Traffic came to a standstill and a significant amount of buildings suffered from flood damages (e.g. ChinaSMACK, 2011; Reuters, 2011). The culturally important Tiananmen Square was flooded though it is protected by the highest standard in Beijing (56mm/hour which corresponds to a 5 year event). Downtown areas are protected are generally protected to withstand a design storm with a return period of 3 years (Vojinović, 2015), while the standards for many other areas are somewhat lower (ibid). One of the major

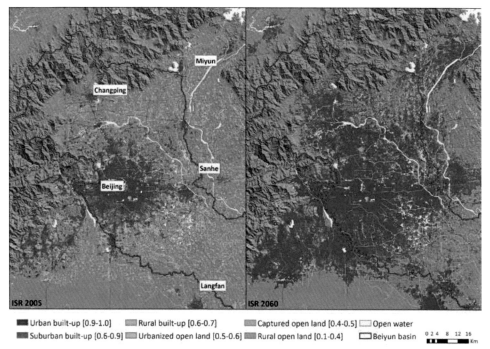

**Figure 29:** Estimated ISR distribution for Beijing in 2005 (left) and 2060 (right)

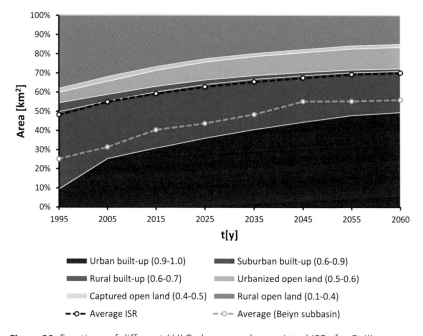

**Figure 30:** Fractions of different LULC-classes and associated ISRs for Beijing

contributors of urban floods is the extraordinary urban growth that took place in the last 20 years (see chapter 4 as well as Appendix A3). Growth containment including the application of a green belt to exclude a buffer around the primary urban cluster had very limited impact (Yang and Jinxing, 2007). To cope with the increasing pressure on the drainage system in combination with the increase in precipitation (Zhai et al 2005), the Chinese government has recently adopted the concept of 'sponge city' (Zhao, 2016) in which widespread application green infrastructure is promoted to better cope with extreme rainfall events. Yet, it remains to be seen how effective this policy to ensure the exclusion of green areas from future development or how the policy will retrofit existing urban areas.

Application and post-processing of the growth scenario outcomes result in a series of maps with LULC classes and associated ISRs. Figure 29 shows the resulting maps for 2005 and the projected horizon of 2060.

Due to the direct relationship between the LULC transitions and the subsequent ISRs, the spatial trends are very similar to those described in section 3.5), describing the projected urban development scenario for Beijing: i) substantial concentric development of the main urban centre into a vast contiguous built-up area ii) suburbs and secondary towns are captured within the main urbanised clusters and iii) the proportion of suburbs and rural built-up areas is declining. The LULC class composition of the urban footprint for the interval 2005-2060 is shown in Figure 30 including the mean ISR which is calculated as the weighted mean of the ISR-fractions. To interpret these outcomes, the results are compared to the projected ISRs for a different extent: the Beiyun subbasin, for which the delineation is depicted in Figure 29. Using a basin or subbasin as an analysis extent is more conventional in rainfall-runoff modelling but is, as already mentioned earlier not necessarily a useful in the case of a megacity where smaller drainage units are more suitable.

The figure clearly illustrates an increase in the ISR over the interval 1990-2060 which is primarily driven by the rapid development of the fraction of urban built-up areas in the in the urban footprint. This fraction increases from only 9.2% in 1995 to currently 30.7% (2015) and finally to an estimated 49.5% in 2060, an increase of increase 40.3% and 18.8% respectively. This increase is mostly at the cost rural open land, for which the proportion decreases by 11.8% in the 2015-2060 interval. Yet, also the relative decrease of suburban built-up areas (-7.9%), which could be already observed in the LULC maps (Figure 29) is clearly visible in the graph. The proportion of urbanized and open land remains more or less stable, although the amount of urbanized open land is marginally increasing. This suggests that a consistent fraction of green urban patches are formed along with the expansion of the city. Thus, the increasing ISR seems to be due to the development of contiguous urban areas: the urbanness of a major portion of the built-up areas has increased and therefore more cells are classified as urban built-up. This in turn causes a shift in the associated ISR values.

Obviously, the projected ISRs are significantly higher than those for the Beiyun sub-basin since the basin extends into the less populated rural areas. In the subbasin the estimated mean ISR for the 2005-2060 interval increases from 31.3% to 56.0% respectively. Furthermore, the growth of the mean ISR seems to flatten out after 2045 which is about 15 years prior to the equilibrium reached in the Beijing urban footprint. Research on the direct relationship between urban development in the Beiyun (sub) basin and subsequent floods due to overland flow has not been found. Research has been performed on flood hazard from the Beiyun River, for which discharge is directly affected by dam operations upstream. Minghong et al (2013) conclude that the flood risk is limited and mostly affected by the operational regime which could be further improved by proper flood forecasting.

For the ISR estimations within the urban footprint, the estimated mean ISR for 2005 is somewhat confirmed by more in-depth studies made by Tan et al (2008), Li et al (2011) and Kuang (2012). Yet, the geographic extents on which these studies are made all differ, making direct comparisons problematic.

The relatively stable fraction of captures and urbanised open land suggests that the fragmentation of built-up areas in Beijing remains relatively stable. Despite the extensive growth, the LULC changes suggest an even proportion of large undeveloped patches of land that could play an important role in providing peak storages or infiltration zones. The estimated FD of the built-up areas provides further evidence for this claim. Although initially dropping during the period 2005-2025, the value remains almost static at a value of about 1.33, which is relatively low. This means that the shape occupied by built-up area is relatively simple. This suggest a relatively compact distribution of built-up areas which can be observed by in Figure 29 for 2060. This is confirmed by assessing the fraction of urbanised and captured open land, which ranges between 17% and 18% for the interval 2015-2060. Compared to most of the other case study areas (see section 6.2.2), this fraction is relatively low.

Although assessing the changes in LULC and associated ISR fractions provide insight characterisation of future composition of the Beijing urban footprint, they do not express the actual growth projected by the scenarios. Especially the growth of built-up areas, which in the case of Beijing signifies the main urban core, has a large impact on the drainage conditions. To express both the changes in proportionality as well as the absolute growth of the main contributors to increasing ISRs, an additional method has been developed, from which the outcomes are shown in Figure 31. The method sets the modelled LULC changes, which are a product of complex spatial interactions (see chapter 3) against a simple, constrained statistical extrapolation of LULC changes. More formally, the extrapolated area $A$ covered by LULC class $i$ at the next interval $(t+1)$ is calculated as follows:

$$(12) \quad A_i(t+1) = \left( \frac{\sum_{i=1}^{N} A_i(t+1)}{\sum_{i=1}^{N} A_i(t)} \right) A_i(t)$$

where $N$ represents the complete set of LULC classes described in Appendix A1. Thus the extrapolated areas can be considered as a baseline from which the modelled results in most cases deviate. This is clear in Figure 31 (left), where the modelled development of urban built-up areas (solid line) far exceeds the extrapolated projections (dashed line). The modelled suburban built-up LULC area diverges from the numerically extrapolated area to become significantly smaller in 2060. This also holds for rural built-up areas. So here it seems clear that the growth of urban built-up areas, which approximate ISRs with 90-100% of soil sealing, is disproportionately high and acts as main contributor of the estimated increase of the mean ISR (see Figure 30). In 2060 the occupied area is estimated about 62% higher than if the growth would have continued by constrained statistical extrapolation. The model outcomes for suburban areas show 27% less growth than the extrapolations would predict. For rural built-up areas this accounts to 56% less growth. The urban built-up areas growth is almost certainly at the costs of suburban and to a lesser extent of rural built-up areas, which seems to be confirmed by the observations of the LUCL distribution (see Figure 29).

On the other hand, Figure 31 (right) also illustrate a disproportionate increase of the the combined captured and urbanised open land. Thus, also the area reserved for GI is growing faster than the statistical extrapolations albeit to a much smaller degree than the urban built-up areas. The subsequent decrease of the ISR is therefore outweighed.

To summarize the projected changes in the drainage performance of Beijing, 3 distinct scale levels can be identified that best represented the presented outcomes:

1. *Macro level*: Mean ISR. This single metric describes the overall drainage of the urban footprint in a single scalar. In Beijing the avg. ISR changes from 0.59 (2015) to 0.70 (2060). At the same time, the total urban footprint increases by 49% to an area of almost 6900 km². Obviously, this area includes the complete Beijing metropolitan area.

2. *Meso level*: FD and fraction of Open Land. A characterisation of the 'urban fabric', i.e. the spatial layout of the built-up areas, is provided by the FD which

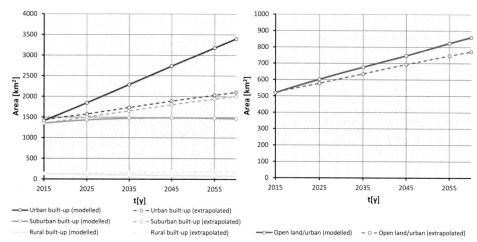

**Figure 31:** Modelled and extrapolated projections of built-up (left) and open areas (right) in Beijing

expresses the fragmentation level of the urban fabric and the fraction of open land (i.e. captured and urbanised open land), which indicate the area occupied by smaller patches and open areas along fringes. Both these metrics indicated on a mesoscale level how distributed space is available for local water storage and/or green infrastructure. Beyond 2015, the FD remains stable around 1.38 which indicates a rather compact composition of built-up areas. Also, the fraction of open land remains stable at around 18%.

3. *Micro level*: Proportionality of built-up area development. By comparing the projected growth produced by the urban growth model to a mere statistical extrapolation, an assessment can be made of the proportionality of the growth of built-up areas. Especially when there is extensive transition from suburban to urban built-up areas, ISRs at local level (raster cell) will increase substantially. For Beijing, the growth of urban built-up area for 2060 is estimated more than 1.5 times the size of a proportionally growth area in 2060.

Although for Beijing all indicators express a negative trend towards future drainage performance, they do not necessarily lead to an overall increase in flood hazard. Nevertheless, future stresses on the existing and planned stormwater drainage network seem to increase disproportionately. This suggests that the standards for future drainage systems need to be increased while possibly, current systems need to be retrofitted to increase their capacity. An alternative is of course to maintain more open spaces to ensure local storage capacity.

The question is how these outcomes compare to some of the other fast growing megacities in the study. This is covered in the next section.

## 6.2.2 Comparing the drainage performance

### 6.2.2.1 Macro level assessment

When comparing the estimated current (2015) avg. ISRs for the respective urban foot-prints, Karachi and Mumbai rank highest (see Table 12) with avg. ISRs of about 62%. The cities lowest ranked are Calcutta and Shanghai with ISRs of 42% and 41% respectively.

For Mumbai this is especially caused by the saturation of the peninsula with a single contiguous urban built-up area. Although constrained by water on the south side, the degree of freedom for development of Karachi is relatively large. Yet, the city is very compact in its current setup: urban blocks are located in vast areas with very little interruption of open land or surface water.

Calcutta shows a high level of fragmentation, and subsequent non-urban built-up areas, which largely influence the overall ISR of the urban footprint. The city's sprawling growth has been identified earlier by for instance Bhatta (2009). The metropolitan area of Shanghai can be considered as a network city in which Shanghai is the largest node. Like the Guangzhou-Shenzhen region, the rapidly urbanising network covers a very large area of over 5000 km2. Yet, in its current state, there are still hosts a relatively large portion of open land in the urban footprint. The resulting avg. ISR is therefore relatively low.

Urban growth shifts the estimated avg. ISRs for all cities to higher values (16.8% on average). In the ranking for 2060, Karachi is still at the top an estimated avg. ISR of 72%,

**Table 12:** Ranking of cities based on ISR for 2015 and 2060

| Rank | City | ISR 2015 | City | ISR 2060 | d2015-2060 |
|------|------|----------|------|----------|------------|
| 1 | Karachi | 0.62 | Karachi | 0.72 | 16.1% |
| 1 | Mumbai | 0.62 | Dhaka | 0.71 | 14.5% |
| 3 | Beijing | 0.59 | Beijing | 0.70 | 18.6% |
| 4 | Dhaka | 0.57 | Istanbul | 0.69 | 21.1% |
| 5 | Istanbul | 0.57 | Mexico City | 0.68 | 27.2% |
| 14 | Jakarta | 0.48 | Delhi | 0.57 | 18.8% |
| 15 | Lahore | 0.45 | Jakarta | 0.53 | 17.8% |
| 16 | Guangzhou-Shenzhen | 0.44 | Lahore | 0.50 | 13.6% |
| 17 | Calcutta | 0.42 | Calcutta | 0.48 | 14.3% |
| 18 | Shanghai | 0.41 | Shanghai | 0.45 | 9.8% |

but is now closely by a number of other cities like Dhaka 71%), Beijing (70%), Istanbul (69%) and Mexico City (68%). Mumbai has moved to rank 6 with an avg. ISR of 67%, which is still within a 5% range of the top-ranked city. At the bottom of the ranking the differences are much larger. Delhi, which is ranked 14 at 57% shows a significant difference with the bottom ranked cities of Calcutta (48%) and Shanghai (45%).

What is more important in the interval 2015-2060 are the projected shifts within the cities. The avg. ISR in Guangzhou-Shenzhen increases for instance by an estimated 31.1% (from 43.6% to 57.2%). Also Manila, Mexico City, Dhaka, Tehran and Istanbul show an increase of more than 20% (see Appendix B2 for the complete table). Mumbai and Ho Chi Minh City on the other hand show only a limited increase of 7.7% and 8.9% respectively.

### 6.2.2.2 Meso level assessment

The estimated FD of the urban footprints differs significantly with a consistently high FD for Manila and Calcutta and a low FD for Delhi, Beijing and Cairo. This confirm some of the earlier made observations about for instance the high level of sprawl in Calcutta and the relative compact urbanised area in Beijing. Yet, also additional information is obtained about the distribution of built-up areas. In Figure 32, the FD is shown for 2015 and 2060. In both years, Manila and Cairo can be considered outliers with a significantly higher and lower FD compared to the bulk of the cities. Manila's current FD is estimated at 1.59 and increases to 1.63 in 2060. The FD of Cairo remains stable at 1.34. When observing the actual LULC maps (see Appendix A3), the reason for the high FD for Manila seems to be produced mainly by the dendrite-like urbanisation patterns in the northern and southern fringe of the metropolitan area. The primary urban cluster, representing the downtown area, is constrained to the east and west by water which causes urban development to saturate the area with urban built-up areas. This would drive the FD down since it produces a significantly more compact urban form. This is also the case for the other outlier: Cairo. Here, urban development seems constrained by the contour of the Nile delta which starts just south of Cairo. As can be observed in produced LULC maps (see Appendix A3), main urban clusters are relatively compact. The compact setup of the Beijing metropolitan area and subsequent low FD has already been described in section 3.3.1 The only cities where a moderate change in FD can be observed in the interval 2015-2060 are:
- *Mexico City*. Decline of the FD signifying a more compact setup;
- *Mumbai, Shanghai and Jakarta*. Increase of the FD which is a sign of increasing complexity and fragmentation of the built-up area distribution.

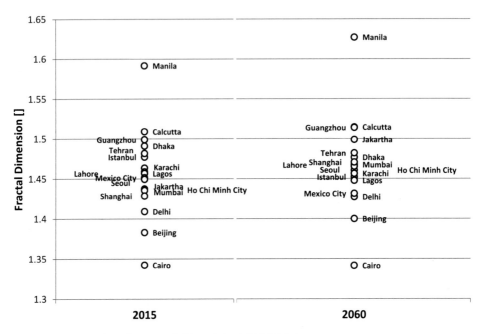

**Figure 32:** Estimated FD for 2015 (left) and the 2060 (right)

A graph showing the FD over the complete interval 2015-2060 is provided in Appendix B3.

From the development of the open land-ratio over the interval 2015-2060, four distinct classes of behaviour can be identified:

- Declining ratios (Figure 33, top-left). For these cities, the fraction of available patches of captured and urbanized land becomes lower over time. Locations for water storage or GI are less frequently excluded from urban development, which increases the pressure on the constructed drainage infrastructure. Cities that fall within this class are Istanbul, Delhi, Lahore, Mexico City, Seoul and Tehran. In Delhi, the decline is most dramatic: 10.3% over the interval 2015-2060.

- Increasing ratios (Figure 33, top-right). Increasing ratios of open land signify relatively more space for water storage and green infrastructure. Urban growth proceeds by leapfrogging or by irregular pattern formation of built-up areas. Cities in this class are Cairo, Ho Chi Minh City, Jakarta and Mumbai. Note that the projected increase over the 45 year interval, is relatively modest for all for cities. In Jakarta the increase is about 4.8%. In other cities the increase is below 3%, which is insignificant considering the period over which

the change is projected.
- Stable ratios (Figure 33, bottom-left). Here the fraction of open land remains nearly equal over time. Cities for which this seems the case are: Beijing, Dhaka and Lagos.
- Irregular ratios (Figure 33, bottom-right). In this class all significantly non-linear behaviour is places. For Manila and Shanghai for instance, the open land fraction initially increases to a peak at around 2035, after which the slowly drops. For Guangzhou-Shenzhen this is also the case, but with a peak around 2030 and a stronger drop in the remaining interval 2030-2060. Calcutta and Karachi show the opposite behaviour: initial sharp drops after which the fraction stabilizes or shows a slight increase.

What ultimately decides for the medium and long term how well the urban fabric is able to absorb or storm water, is the combination of a high open land fraction combined with a high FD. Manila is is the highest ranked city for both these metrics: a FD of around 1.6, which exceeds the complexity of the coastline of Norway (Feder, 1988) and an fraction of open land that projected to exceed 27% beyond 2035. Other cities that perform well are for instance Jakarta and Shanghai. The characteristics of urban fabric of Guangzhou-Shenzhen and Calcutta seem to converge to each other: in 2060 their FD and open land fraction are almost the same. Low performing cities are especially Cairo, with the lowest FD as well as open land fraction below 14%. Other low scoring cities are Beijing and Mexico City, where Beijing is scoring low but stable while Mexico City's performance seems to decline rapidly over time.

### 6.2.2.3 Micro level assessment

As described in the section covering Beijing case study (6.2.1), it is important to evaluate the proportionality of the projected urbanisation against a mere statistical extrapolation of current LULC fractions and associated ISRs. From the assessment of the proportionality all cities in this research, three different classes can be identified:

1. *Densification*. This is characterised by a disproportionate increase of urban built-up areas at the expense of suburban and rural built-up areas. The analysis outcomes for Beijing (see Figure 31, left) are a typical example of this class, where the overall composition of the urban footprint shows an increasing fraction of urban built-up areas representing raster cells with high ISRs (0.9-1.0). This increase is composed of urban expansion but also of replacement of existing suburban and rural built-up areas (i.e. village centres). This class of behav-

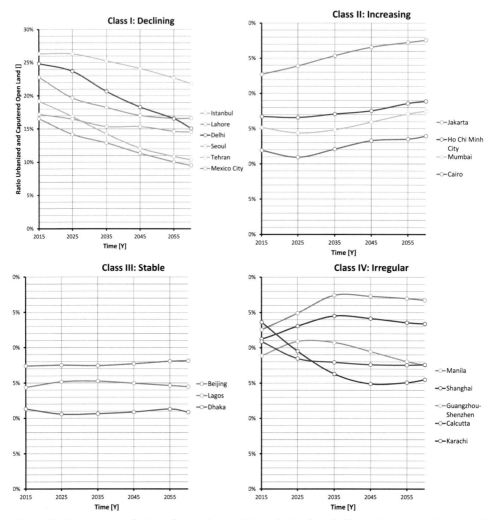

**Figure 33:** Progression of ratio of open land with declining (top-left) and increasing (top-right) trends as well as stable (bottom-left) and irregular (bottom-right) trends.

iour might have the biggest impact on the drainage capacity at local level. Apart from Beijing, cities that fall within this class are: Cairo, Dhaka, Ho Chi Minh City, Manila and to a lesser extent: Jakarta, Karachi and Lagos.

2. *Expansion*. The urban footprint shows a disproportionate growth of urban and suburban built-up areas. This behaviour is more ambiguous, since it depends on the division of growth rates between those two classes. Note that this class does not necessarily rules out substitution of suburban to urban built-up areas. Yet, these transitions occur probably at a lower rate than for class 1. Cities that fall within this class are: Calcutta, Delhi, Guangzhou-Shenzhen and Tehran.

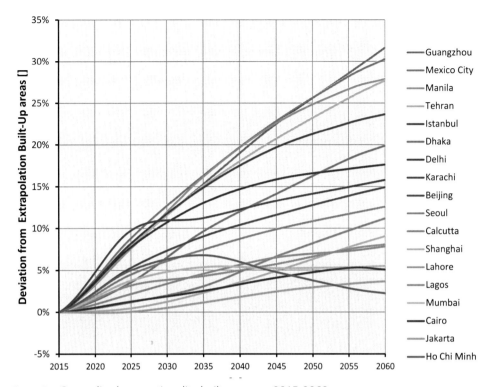

**Figure 34:** Generalized proportionality built-up areas 2015-2060

3. *Urban expansion*. This behaviour can be regarded as a hybrid between class 1 and 2, showing disproportionate development of urban built-up areas and a proportional development of suburban built-up areas. Patches of rural built-up area are, as in the two other classes, declining. Like for Densification, the net outcome of this class is larger fraction of urban built-up areas, albeit at lower rate. Cities in this class are: Istanbul, Mexico City, Shanghai and Seoul.

Obviously, other classes of behaviour are conceivable involving for instance a dispro-portionate decline of urban built-up areas. Yet, these did not occur for the cities used in this study. The outcomes per city are shown in Appendix B4 but summarized in Fig-ure 34, where the weighted disproportionateity for the three classes for built-up areas are shown. Note that the order in the legend is based on the rank in 2060, so in 2060 Guangzhou-Shenzhen shows the highest level of disproportionate growth and Ho Chi Minh City the lowest.

As already indicated, all cities show a disproportionate increase of built-up areas and associated ISRs, which is clear from Figure 34 where no line moves below the 0% mark indicating proportional growth. Only Jakarta seems to show signs of proportional growth until 2025 after which the projections move somewhat above the statistical extrapolation. Contrary to that, the Guangzhou-Shenzhen area as well as Mexico City, Manila and Tehran consistently move towards disproportionate increase of built-up areas about 28% or more in 2060. While for most cities, the proportionality behaves rather consistent, the behaviour of Karachi and Ho Chi Minh City shows initially a rapid increase in disproportionate growth of ISRs while slowing down or in case of Ho Chi Minh City, even dropping after 2030 and 2035 respectively. This suggests that after an initial abundance of suitable land to absorb the urbanisation pressure, this availability become more constrained. Additionally, the number of suburban and rural built-up patches for transforming into urban built-up areas, dropped after initially being available in large quantities. Finally, some additional behaviour can be identified for instance Mumbai, for which disproportionate increase of built-up areas flatten out beyond 2035. For Shanghai, disproportionate growth seems to be accelerating.

What seems counterintuitive, is that the behaviour of the top ranked cities in Figure 34 in regards to their proportionality covers all 3 classes described earlier: Manila's growth is assigned to class 1 (Densification), Guangzhou-Shenzhen and Tehran to class 2 (Expansion), and Mexico City to class 3 (Urban Expansion). This suggests that the actual growth rate compared to the extrapolation of especially urban built-up areas dominates the behaviour of the classes representing suburban and rural built-up areas. Disproportionate development of suburban and rural built-up areas becomes untraceable, in the weighted mean that is depicted in Figure 34. Nevertheless, to explain the interaction between the projected growth of these classes, the proportionality of all three classes need to be examined, as depicted in Figure 31. These are provided in Appendix B4.

## 6.3 Outcomes: Towards a Sponge City

The previously described metrics attempt to provide a characterisation of the urban fabric and the potential implications for the drainage performance as a function of urban growth. Integrating these outcomes into a single appraisal of the projected future drainage performance is not straightforward, especially since the applied metrics cover different scale levels. Only a qualitative appraisal can be applied to create a more schematised outlook. Ultimately, local conditions define which of the four metrics at the three scale levels provides most insight. For instance, the relative high standards and implementation of the constructed drainage network in Beijing allow for less emphasis on both macro- and microscale assessment outcomes. Construction of highly engineered drainage systems with a will likely ensure proper performance and limit stress on the existing overall system, i.e. at the level of the complete urban footprint. Yet, Beijing might be required to focus on the mesoscale level, which attempts to provide information about the distribution of open areas in case of infrequent conditions leading to exceedance of the drainage capacity. In other cities, where it is highly unlikely that investments in drainage can keep up with the rapid development, like Dhaka and Lahore, the macro- and mesoscale might dominate the future outlook. Nevertheless, an overall qualitative evaluation is provided for the purpose of a general summary. This is provided in Table 13, where the different indicator values have been awarded to qualitative categories ranging from 'very negative' to 'very positive' in relation to sustaining the existing urban drainage capacity. The cumulative scores of the three scales, in the range [-2,2] define the overall outcome in the first column. Note that the threshold values on which this classification is based are provided in Appendix B1. Since the aim of this study is primarily to assess and compare the relative changes that

Table 13: Qualitative assessment of the cumulative and scale dependent indicators

| | Cum-mul. | Macro dAvg. ISR | Meso dFD | Meso dOpen Land | Micro Dis-prop. | |
|---|---|---|---|---|---|---|
| Beijng | | | | | | Very positive |
| Cairo | | | | | | Positive |
| Calcutta | | | | | | Neutral |
| Delhi | | | | | | Negative |
| Dhaka | | | | | | Very negative |
| Guangzhou | | | | | | |
| Ho Chi Minh | | | | | | |
| Istanbul | | | | | | |
| Jakarta | | | | | | |
| Karachi | | | | | | |
| Lagos | | | | | | |
| Lahore | | | | | | |
| Manila | | | | | | |
| Mexcio City | | | | | | |
| Mumbai | | | | | | |
| Seoul | | | | | | |
| Shanghai | | | | | | |
| Tehran | | | | | | |

occur due to projected urban growth scenarios, the factors have been evaluated on their expected rate of change during the interval 2015-2060. The actual boundaries used for the qualitative classes are based on the statistical distribution, rather than on accepted thresholds. This is simply since thresholds do not exist and depend highly on local conditions. This issue is covered in more detail in the discussion. Note that for obtaining the qualifications at mesoscale, the two factors have been averages.

When evaluating the overall changes in projected performance, three cities stand out: Ho Chi Minh City, Jakarta and Mumbai. Ho Chi Minh seems to perform well at meso-scale while neutral at macro- and microscale. When evaluating the actual distribution of the projected LULC changes (Appendix A3), it seems that the dendrite-like patterns of built-up areas along major roads give rise to a relatively porous urban fabric. Yet, the main urban core seems to be further densifying so abundant levels of open land are not distributed uniformly over the urban footprint. Jakarta seems to be perform-ing even better at a meso-scale level. Also here, the favourable changes in drainage conditions at mesoscale can be mostly contributed to the fragmented expansion along the east-west axis of the city. In Mumbai the conditions are different: the constrained space available for urban expansion on the peninsula gives rise to leapfrogging and ex-pansion of suburbs that are relatively disconnected and remote. So favourable changes in future drainage conditions at mesoscale are mostly due to the characteristics of Mumbai's peripheral cities like Navi Mumbai, Dombivli, Bhiwandi and others. Thus, in a way, the projected development of Mumbai is the opposite of spatial trends occurring in the Guangzhou-Shenzhen area or in Shanghai where an existing network of smaller cities is densifying.

Cities scoring particularly low are Istanbul and Karachi where all contributing factors have a negative or very negative outlook. Closer inspection of the maps showing the projected LULC changes, reveals for Istanbul a very compact and constrained develop-ment in both the European and the Asian part of the city. Karachi, as already described shows a compact but substantial development which adversely affects drainage per-formance at micro-, meso- and macroscale.

## 6.4 Discussion

The produced comparison, using a set of indicators on macro-, meso- and microscale does not allow for a straightforward evaluation. All serve as proxy indicators represent-ing a more complex phenomenon that can only be assessed by much more detailed ap-proaches on individual case study level. A multitude of biophysical and socio technical conditions like rainfall characteristics, current and future drainage standards and management practises, groundwater and soil conditions as well as building characteristics at plot level ultimately define how cities can cope with rainfall. For instance, drainage conditions in Ho Chi Minh City, Jakarta and some of the other cities are severely impaired by subsidence (e.g. Phi, 2007; Abidin et al, 2011) due to groundwater extraction and building activities. If anything, the outcomes merely attempt to provide clues on how the conditions change compared to the current conditions. In Beijing for instance, where rainfall induced local floods are rare compared to Dhaka, Karachi or Lahore the implications of future urban development are very different from compared to a city where during the monsoon the streets are blocked several times a week due to excessive rainfall. In turn though, the consequences of impoverished drainage conditions in those types of cities, might ultimately lead to impaired living conditions, sub economic performance or during extreme events to severe economic and health impacts.

Obviously, the indicator set used to evaluate the drainage performance is dependent on the available datasets. A basic requirement throughout this study was that these are publicly available and provide global coverage in order any city can be included in this assessment. Aside-effect of this requirement is the limited expressiveness achieved in the datasets produced in the projections. Given the 30m resolution of Landsat TM and ETM+ images, a precise characterisation of building footprints, infrastructure and other classes of built-up areas was not possible. This problem extends to the development of ISRs where building typologies on parcel level might have a significant impact on the projected outcomes. Future ISRs in high density urban areas might therefore be lower instead of higher. In Beijing for instance, DHI (2009) concluded that in many urban redevelopment areas, the ISR might actually increase due to the construction of high-rise apartments surrounded by green areas. The datasets used for the projects cannot express such changes effectively also because they sometimes appear at subpixel level. Note that due to increasing car ownership, over time such green zones often turn into parking lots which undo the potential drainage advantage. Ultimately, one can ask though if such a level of detail is required in the case of strategic urban development projections for megacities. Further research is required to compare model outcomes using a limited and broad dataset as model input.

## 6.4.1 Policy options

Do the presented outcomes provide clues for future policy options, strategic interventions or even measures? One argument is that the division in scale levels where the assessment is based on also provides different opportunities for intervention. The overall increase of the avg. ISR of the urban footprint for instance supports need for the excluding large patches of land from development. In a way, such a microscale approach might lead to a very concrete intervention. Obviously, actual identification of such areas is dependent on the local conditions and needs to be further investigated. The urban development of many of the cities is constrained by major rivers. In many cases, the past urbanisation of floodplains (see section 4.4.1) already saturated the possibilities for urban growth. In some cases this lead to an increase of drainage problems since construction of river dikes and elevating adjacent land caused drainage congestion; runoff was no longer able to flow into the rivers which served as drains. This trade-off is especially visible in Dhaka, where embankment of the western part of the city has led to increased waterlogging of urban areas (Mark et al, 2001).

The mesoscale level mostly focusses at the characteristics of the urban fabric; is open land available for green infrastructure or water storage at regular intervals? Such an appraisal might provide new or additional requirements to urban development or re-development projects covering neighbourhoods or several blocks of buildings. In a way, such requirements increase the fragmentation of urban built-up areas to include more patches of 'urban green and blue'. Yet, using urban fragmentation as a catalyst for retaining urban drainage capacity is only useful when that level of fragmentation is maintained in the future. If carefully planned open spaces end up as parking lots or commercial properties, the impact of the adapted urban designs is lost. Finally, changes in the building codes might have a direct impact on decreasing the ISR at plot level, where extensive building footprints prevent drainage at local level. Also, preventing private gardens and courtyards from being covered by pavement is part of such an approach.

Yet, in many of the cities that have been included in this study effective zoning plans, enforced building codes as well 'smart urban design' that includes GI is often a luxury. Yet, providing a strategic long term outlook for such cities might act as a signpost that increases the urgency for acknowledging the potential future impacts of urban growth on urban drainage.

## 6.5 Conclusions

The assessment of the future drainage performance of fast growing megacities as a function of projected LULC changes is not a straightforward task. Urban drainage depends on a complex interaction of rainfall, soil characteristics, spatial distribution of built-up areas as well as associated manmade drainage structures. Subsequent flooding and impacts depend on even more complex factors, including detailed information about the exposed assets. Typically, assessments at strategic level focus on ISR and their spatial distribution. At a detailed scale of neighbourhoods, 1d-2d coupled hydraulic models are typically require extensive data for setup and calibration which do not fit with the expressiveness of LULC change scenarios. In this study an alternative approach is applied that attempts to fully exploit the limited information derived from the LULC scenarios at different scale levels. Focussing on the changes in the individual growth scenarios that could affect the future drainage performance, evaluations have been made for all cities at macro-, meso- and microscale. The outcomes are presented as relative changes compared to the current conditions. On macroscale level the focus is on the mean ISR in the urban footprint. At mesoscale level, the focus is on the FD and the open land-fraction which attempt to measure the distribution of green areas over the urban fabric. Finally, at the microscale an assessment is made on the proportionality of the development of urban, suburban and rural built-up areas which are associated to different levels of ISRs. All assessments are performed over the period 2015-2060, which covers the horizon of the LULC projections.

The outcomes at macroscale show that Karachi, Mumbai and Beijing are projected to have the highest ISR-levels in 2060, exceeding 70% of paved surfaces. Yet, the cities that are expected to show the largest changes in ISR are the Guangzhou-Shenzhen area as well as Manila, Mexico City and Dhaka where ISRs increase by 25% or more compared to the 2015 values. Cities that do relatively well are Ho Chi Minh City and Mumbai although also here, the avg. ISR is increasing.

At mesoscale level, the relatively worst performing cities are Istanbul and Karachi where the urban fabric seems to become much more compact, the fringes less fragmented and the overall availability of open land for water storage or infiltration becomes increasingly sparse. Jakarta, Mumbai and Shanghai on the other hand perform relatively well. Although they all have very different growth characteristics, the porousness of the urban fabric shows a relative improvement.

At microscale level all cities show a disproportionate growth of high density built-up areas. This happens sometimes at the expense of suburbs or villages and sometimes rural areas are simply directly transformed into urbanised land. Here, Guangzhou-Shenzhen, Mexico City, Manila and Tehran transition to high density urban areas more rapidly than if the projected urban growth would be statistically extrapolated.

When the outcomes of the different scale levels are combined, the best and worst performing cities are Jakarta, Ho Chi Minh City, Mumbai (best) and Guangzhou-Shenzhen, Karachi, Istanbul, Mexico-City and Tehran. Focussing on the relative changes, these outcomes become relevant when compared to the current drainage performance. So for cities, that currently experience little or no urban floods might in the future experience nuisance flooding with small associated impacts. Yet, cities where severe monsoon driven urban flooding occurs on a frequent basis might face significant future consequences policies and investments in urban drainage do not keep up with the projected urban growth.

# 7. Adding depth: Estimating flood damages in Dhaka

This chapter is based on:

Khan, D. M., Veerbeek, W., Chen, A. S., Hammond, M. J., Islam, F., Pervin, I., ... & Butler, D. (2016). Back to the future: assessing the damage of 2004 DHAKA FLOOD in the 2050 urban environment. Journal of Flood Risk Management.

## 7.1 Introduction

In developing countries flood damages are often relatively high due to absence of sufficient flood protection and inadequate infrastructure. Furthermore, low-lying lands and drainage routes are being encroached upon to build settlements. This is also the case in Dhaka: a megacity facing an unprecedented urban growth combined with a history of major flood events. Bangladesh has been subjected to catastrophic flooding for many years. Since 1950, 16 catastrophic floods (Ahmed, 2014) have occurred from which the 1955 and 1974 floods were most devastating. In more recent years catastrophic floods occurred in 1988, submerging more than two thirds of Bangladesh between August and September and in 2004, flooding about 38% of the country and affecting more than 25% of the total population.

Since the city is located adjacent to the Buriganga, Turag and Balu rivers, Dhaka has been subject to periodic flooding since its early days. The recent 1988 flood affected around 30% of all (861 thousand) residential buildings causing damage of about Tk. 2.3 billion (about 122 million 2013 USD). Furthermore, about 384 km of paved roads were inundated during the event as well as water supplies from deep tube wells (Alam and Rabbani, 2007). During the nationwide floods in 1998, caused by extreme river levels, 23% of the area of the western city was inundated. Even so, the western side was less affected compared to the unembanked eastern side due to the flood protection measures (Siddiqui and Hossain, 2006).

One of the major obstacles for comprehensive research into Dhaka's floods risk is the poor understanding of design floods. This especially holds for riverine flooding due to the complexity and volatility of the Bangladesh river system, located in the largest delta of the world (Mikhailov and Dotsenko, 2006). The absence of design floods prevents a structured flood risk management approach based on exceedance probabilities, which as a consequence is limited to an event driven approach. This also has affects the assessment of urban growth induced future flood risk: the link between flood probabilities and associated impacts cannot be determined for a future urban extent that is substantially larger than the current extent. To overcome this issue, a more practical approach has been developed where a historic flood event has been projected onto a future projection of the city.

The catastrophic flood from 2004 was chosen for such an approach since relatively much information on impacts were available compared to for instance the even larger flood of 1988. The flood arose from a combination of extreme river flows and intense rainfall in the early months of the monsoon, between June and July 2004.

In this chapter the previously investigated consequences of urban growth on the flood extent are extended by an attempt to assess the changes in flood consequences. After a further introduction of the flood related challenges Dhaka is currently facing, the respective models, growth scenario and outcomes are presented that attempt to provide insight in the question: "What would be the impact, if the flood that hit Dhaka in 2004, would occur in 2050?". The focus of the chapter is twofold: (i) estimating the possible increases in flooding as a consequence of extensive urbanisation and (ii) the potential increase in flood damages given the rapid urbanisation, through increased runoff and exposure of assets.

## 7.2 Dhaka case study

### 7.2.1 General characteristics of the city

Currently hosting around 14 million inhabitants, with annual growth rates exceeding 4% (Demographia, 2013) which are among the highest in the world, Dhaka is expected to grow to the third largest metropolitan area by 2020 (UN, 2011). Apart from being the political capital of Bangladesh, Dhaka also serves as the economic capital producing up to a third of the total GDP of Bangladesh. Although Dhaka is a megacity, the urban footprint is relatively small. The city itself covers only about 400 km² and is therefore one of the most densely populated cities in the world. In some areas, the density exceeds 80,000 people per km² (See Appendix A3). The city has historically attracted many migrant workers. Half of the workforce is employed in household and unorganised labour, whereas about 800,000 work in the textile industry.

After the 1988 flood a Western embankment was constructed protecting the most populated areas of Dhaka city (see Figure 35). Yet, major parts in the eastern area are still unprotected while conversion of agricultural land into built-up areas is proceeding rapidly. Apart from exposing a growing number of people and assets to river floods, the development further increased the problem of waterlogging since drainage capacity is significantly reduced. This is already a considerable problem in western Dhaka where due to limited pumping capacity, increased monsoon driven waterlogging occurs. Encroachment and unauthorized land-filling further decrease the drainage capacity of the city. Enforcement of zoning laws as well as the 'Wetland Preservation Act' have not been effectively enforced (Rahman et al, 2005).

The risks associated with floods are expected to increase in coming years due to high rates of urbanisation. The scale and condition of the drainage system has not kept pace with the rapid urban expansion and development. Dhaka can be divided into two: the western and eastern areas (Figure 35). The western part of the city is protected by embankments and has a storm sewer system. The eastern part is the lowest lying part of Dhaka, and faces the most severe risk of flooding. The area is not protected by embankments and mainly consists of open channels. Because of the land scarcity in the city, the population of the eastern fringe has increased rapidly during the last decades. Thousands of people have encroached upon low-lying areas in search of a place to live, reducing the efficiency of natural drainage, and increasing the flood risk.

## 7.2.2 Urban growth

Dhaka has experienced significant urban growth during the last 50 years (Dewan and Yamaguchi, 2009). Urban development and economic growth have sparked a construction boom; high-rise buildings and skyscrapers have changed the city landscape. Between 2000 and 2011, 1.8% of the urban growth was due to infill. During that period, urban extension and leapfrogging were estimated at 82% and 16.2%, respectively (Dewan and Corner, 2014), which suggests that urban development is mostly occurring along the fringes of Dhaka. Most of this extension took place on the northeastern part of the city (Dewan and Corner, 2014) as well as on the southeastern part in the Dhaka-Narayanganj-Demra triangle. In the southwest, urban development is heavily constrained by the rivers, which provide an extensive barrier (Figure 36). The eastern

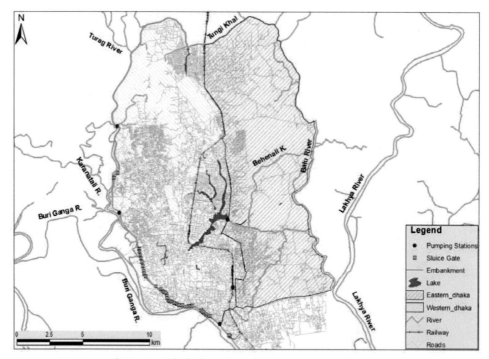

**Figure 35:** Eastern and Western Dhaka based on drainage separation

side of Dhaka is facing a substantial amount of land grabbing, where block-by-block development is occurring in the lower lying areas. Leapfrogging development also occurs in the Dhaka metropolitan regions. This is especially the case in areas like Tongi, Gazipur and Savar. These regions might become more prominent in the future, as population continues to grow. Figure 36 shows the urban development within the low-lying areas from 1990 to 2005, based on Landsat ETM+ derived LULC maps. The continuing unplanned and unmanaged urban development has a severe impact on the LULC, which consequently impacts the physical characteristics of the flood regime. Although severely limiting the infiltration capacity that gives rise to seasonal drainage

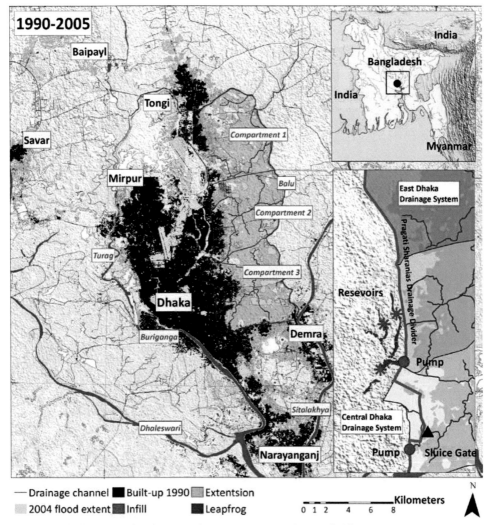

**Figure 36:** Dhaka urban development between 1990 and 2005 (left) and eastern Dhaka drainage system (right).

congestion, the ever-increasing concentration of people and assets in flood-prone areas significantly increases the potential consequences of floods.

### 7.2.3 The 2004 flood

The catastrophic flood that hit Dhaka in 2004 (see Figure 36), caused major health impacts and economic losses (Siddiqui and Hossain, 2006). First, high river discharges caused by prolonged monsoon rainfall affected much of eastern Dhaka. In September 2004, extreme rainfall over 4 days overwhelmed the capacity of the urban drainage system which subsequently led to extensive flooding in the protected area of western Dhaka as well (Alam and Rabbani, 2007). A major flood also occurred in 2007 that significantly affected the eastern part. Untreated waste water mixing with flood water caused a large number of patients getting admitted for water-borne diseases (Islam et al., 2008).

The 2004 flood was estimated to have a return period of 100 years (Rahman et al., 2005). The subsequent flood damages of the 2004 flood were estimated by Islam (2006) and (Hossain, 2006) and ranged between USD2.2 and USD6.6 billion respectively. Hasnat (2006) urged for future assessments and provided recommendations and a procedure for the task.

## 7.3 Urban growth model, flood model and damage model

### 7.3.1 Flood model

The hydraulic modelling for Dhaka comprised two separate models to reflect the unprotected eastern side with mostly open channels and protected western side with mostly piped drainage. The eastern side was delineated into three compartments or hydrological units that were used to calculate the surface runoff resulting from the changes in LULC. The modelling relies upon hydrological inflows at certain points as well as precipitation inputs. An urban rainfall–runoff model, based on the time–area method, was used to estimate the runoff from the rainfall. The central part was divided hydraulically into 22 subcatchments, connected by a single primary drainage system. The primary line is divided into secondary, tertiary as well as a catchpit connection system. To prevent backflow towards the city during the monsoon period when the water level of the Balu River is higher, there are three drainage control structures that were included in the model. The cross sections of the open channels in the eastern model were updated based on 2004 conditions.

The hydraulic model of the drainage network was then coupled with the digital elevation model (DEM) with a cell size of 25m to simulate two-dimensional (2D) surface runoff propagations. Mike Flood was used to perform the modelling for both the one-dimensional drainage and 2D surface models. In the 2D model, the retention ponds were replaced by the pond bathymetry and surrounding connectivity was established accordingly. In central Dhaka, most of the urbanised areas are at elevation of 6–8 m above the mean sea level. Conversely, more than 70% of the area of Eastern Dhaka is below 6 m.

## 7.3.2 Flood damage model

Although the requirements and methods to perform a comprehensive flood damage assessment that include direct and indirect damages are well established (e.g. Penning-Rowsell et al., 2005; Messner et al., 2007; Merz et al., 2010), operationalising such a task in the context megacities in the developing world is virtually impossible due to the extensive data requirements across many economic sectors and assets. The development of a flood damage model and associated depth damage curves (DDCs) that relate expected damages to inundation depths have therefore been limited to residential and commercial properties only (Hammond et al., 2012). DDCs for Bangladesh have been developed previously by Islam (2005) but were updated for the COR-FU project using data acquired by a survey among 215 households and 215 businesses that suffered flood damages in recent years (Haque et al., 2014). In order to be used in combination with the urban growth scenarios, the property-based (i.e. building type) DDCs were aggregated and converted to fit with the corresponding 10 LULC-densi-

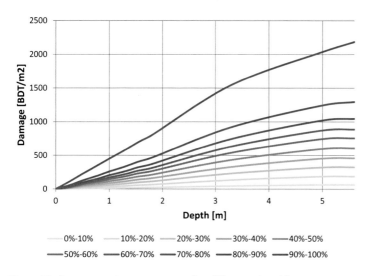

**Figure 37:** Aggregate damage curves for different densities

ty classes. Although some information from the DDCs is lost during aggregation, this step is essential when integrating the future LULC scenarios in the damage modelling. The resulting DDCs are depicted in Figure 37. Note that for inundation levels beyond 5.6 m, the damage levels are kept at the depicted maximum level for the respective LULC-density class.

The actual damage calculations were performed by (i) determining the inundation depth for each individual built-up LULC patch from a given flood depth map; (ii) identifying the associated damage level using the lookup table from the appropriate DDC and (iii) aggregating the damage of patches with the same LULC. These steps were combined in a single model developed by Chen et al. (2015) that also compensated for potential misalignment of the spatial characteristics of the required flood maps (e.g. regular grids, irregular meshes) and feature maps (e.g. LULC grids, polygons).

## 7.4 Scenarios

For this study, growth scenarios were developed using the model introduced in chapter 3. In extension to the basic setup of the model, for this application a more extensive set of base maps were used. Instead of the division of urban areas into 3 different LULC classes (i.e. low, medium and high density built-up areas), built-up areas were divided into 10 different density-based classes. Instead of using only LULC maps based on Landsat remote sensing data as is the case for the case study areas throughout this study, additional data sources were used. Especially the Detailed Area Plan (DAP)

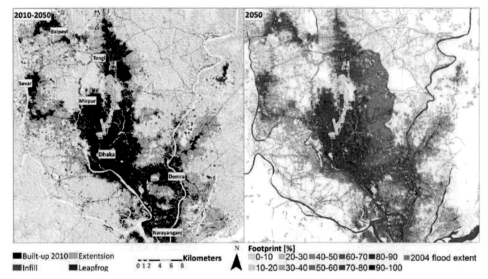

**Figure 38:** Growth characterisation (left) and resulting distribution of built-up areas including flood extent (right).

which provides a detailed (vector-based) map of all structures in Dhaka, proved to be useful to further differentiate between the footprint of built-up areas.

In total a number of nine urban development scenarios were considered, which were combinations of three population growth and economic development conditions, developed by Schlitte (2013), and three spatial development scenarios that include a BAU-based extrapolation of past development trends (Veerbeek, 2013). The spatial consequences of the alternate urban growth rates were translated into the growth model as alternate transition rates, thus increasing the number of cell transitions to urban built-up areas and between built-up areas. Out of these nine scenarios, the high growth rate scenario resulted in the most dramatic LULC changes in flood-prone areas. Consequently, this scenario was regarded as the upper bound for the potential flood damages as a function from the changing distributions in built-up areas. This LULC scenario was therefore used in the hydraulic model as well as damage model to compare the estimated flood damages for the long-term horizon of 2050. The growth characterisation as well as the LULC distribution is shown in Figure 38.

The three scenarios considered are:

- Scenario 1: Baseline condition, flood damage in study area in 2004;
- Scenario 2: Projected flood damage in 2050 with a projected 250 urban extent but without considering the impact of urban growth on hydraulic conditions;
- Scenario 3: Flood damage in 2050 with a projected 250 urban extent but with the impact of urban growth on hydraulic conditions

The 2050 BAU scenario urban growth is characterised (see Figure 38) by 7.3% infill, 84.6% urban expansion and 8.1% of leapfrogging. Compared to observed recent development of Dhaka (i.e. 1995–2005), the projections suggest an increase of urban infill and a reduction of leapfrogging development. This observation is supported by measuring the fractal dimension (FD) of the built-up areas, which provides a measure of the complexity of the observed perimeter in relation to its footprint (e.g. Turner, 1990; Herold et al., 2005). The FD dropped from 1.51 in 2005 to 1.47 in 2050, indicating a more compact urban form with less open areas or complex perimeters.

The hydraulic modelling focused on the particular conditions of the 2004 event. Nearly 600 mm of rainfall occurred in a 5-day period during September 2004. This rainfall was included in the model. The water level in the Balu River during 2004 was used as the boundary condition for the hydraulic model. The mean ISR in the Dhaka metropolitan area as a result of the changing urbanisation almost doubles from 0.26 in 2004 to an estimated 0.43 in 2050, compared to 0.36 to 0.54 increase in the unembanked eastern

**Table 14:** Model combinations for scenarios

| Scenario # | UGM | Flood Model | Damage Model |
|---|---|---|---|
| 1 | 2004 Condition | 2004 Hydrology | 2004 LULC |
| 2 | 2050 BAU | 2004 Hydrology | 2050 LULC |
| 3 | 2050 BAU | 2050 Hydrology | 2050 LULC |

part of the city. The changes in imperviousness and subsequent runoff were incorporated in the setup of the flood model. Scenario 1 basically mimics the 2004 conditions, i.e. the reference scenario. Scenario 2 adds the growth projection for 2050 while scenario 3 also includes the subsequent changes for the overland flood conditions due to LULC changes and associated changes in ISRs. Table 14 shows the combination of models used for each scenario.

## 7.5 Outcomes

The produced flood and damage maps are presented in Figure 39. For reference, the DEM clearly shows the relatively low-lying area of eastern Dhaka, which almost coincides with the flood extent. The estimated inundation depths are considerable; flood levels reach over 6 m in some of the uninhabited areas. Yet, even in medium-density

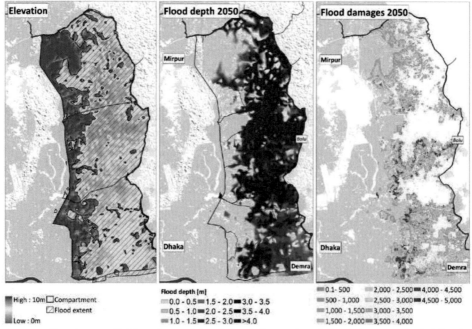

**Figure 39:** Terrain map (left), estimated flood depth (centre) and flood damages (right) for the 2050 scenario.

built-up areas, flood level often reach several meters.

## 7.5.1 Comparison of flooding

During the 2004 event, about 80 km2 of the study area was inundated (almost 65%) of which the majority was located on the eastern part. In the 2050 growth-induced condition, the increasing ISRs have a substantial effect on the estimated runoff volumes and peak levels in the three compartments of eastern Dhaka. These are shown in Figure 40. For all three compartments, the model outcomes clearly showed a dramatic increase of peak flow rates, which were estimated 6.3, 6.1 and 4.5 times higher than those in 2004 for compartments 1, 2 and 3, respectively.

Despite the substantial increase in peak and total runoff volumes, the estimated effect on the actual flood seemed to be minimal. This is because flooding in eastern Dhaka was mainly caused by overtopping of the Balu River banks, with a flood volume that far exceeded the impact of any changes in runoff volume. Consequently, the model outcomes produced only limited changes in flood extent and depths. In the baseline condition (2004), the total flooded area in eastern Dhaka was 70.7%, whereas in 2050 the flooded area was estimated at 71.5%. The changes in expected flood depths are somewhat more significant. Here, a trend could be observed towards higher inundation depths for the 2050 scenario. The total area with expected inundation depths of 3 m or more increased at the expense of areas previously flooded with inundation depths between 1 and 3 m.

Likewise, little changes in inundation were observed in central protected part of the city. This was because the flooding occurs due to drainage blockage at the outfall. In most areas though, the network had adequate capacity. In Central Dhaka about 87% of the model area was inundated up to 0.5 m in the baseline condition whereas in 2050, 92% of the area was expected to be inundated up to the same depth.

## 7.5.2 Comparison of flood damages

The aggregate damage for the 2004 event was estimated at USD 22.8 million for the

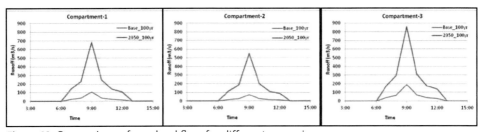

**Figure 40:** Comparison of overland flow for different scenarios

case study area. For the 2050 BAU high growth scenario the damages were expected to increase almost sevenfold to USD 154.8 million. If the impact of urban growth on the hydrologic conditions is also considered in the flood modelling, the damage would further increase to USD 178.1 million for the same event. The comparison of the estimated damage from the different scenarios is shown in Table 15.

Figure 41 depicts the relation between the estimated flood damages and the 10 LULC-density classes. To provide a qualitative classification, the densities have been designated as open-, low-, medium-and high-intensity developed areas using the standardised classification scheme by Homer et al. (2004) used elsewhere in this study. In the baseline scenario, the majority (70.4%) of the area was occupied by either open- or low-intensity developed areas that together suffered about half of the total flood damages (55.6%). In the development scenario for 2050, this proportion shrunk to almost half the original size (38.3%) while the estimated share of damages lowered to 36.4%. Although high-intensity developed areas increase more than threefold in size, the share of damages almost remains the same (7.4%–8.3%, respectively). Subsequently, the most dramatic shift occurs in the medium-intensity developed areas that double in size and suffering more than half (55.3%) of all flood damages.

These figures show flood damages that were formerly suffered in mainly rural and peri-urban areas, characterised by detached farmhouses in small villages, shift to medium-to high-intensity urban areas with urban footprints of up to 80% of ground cover. This also means that the spatial distribution of flood damages is changing, from more dispersed small pockets to more homogenous residential areas.

## 7.6 Interpretation

The first and foremost outcome of this study was the disproportionate increase of expected flood damages as a function of extensive urban growth in eastern Dhaka. Although the growth scenario projected a 49.5% increase of built-up areas in eastern

**Table 15:** Total damage in study area for different scenarios

| Scenario No. | Scenarios | Damage (million USD) |
|:---:|---|:---:|
| 1 | Reference scenario 2004 | 22.8 |
| 2 | 2050 BAU high growth scenario | 154.8 |
| 3 | 2050 BAU high growth scenario and subsequent hydrologic condition | 178.1 |

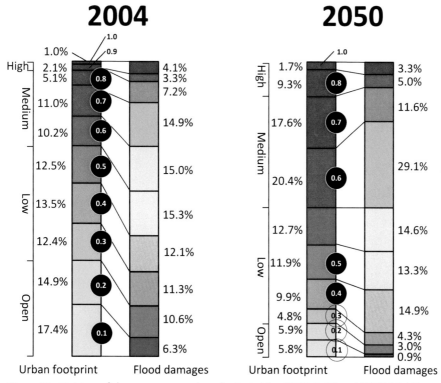

**Figure 41:** Division of damages over urban footprint for 2004 (left) and 2050 (right).

Dhaka (compared with 70% in the Dhaka metropolitan area), the flood damages increased almost eightfold. Thus, both in absolute and relative terms future flood damages increased considerably. Even if the hydrological characteristics of the projected flood are kept the same, the increased exposure of urbanised areas to the flood could be subject to massive impacts that might take a substantial recovery period.

These outcomes were of the same order of magnitude as the estimated increase of the urban flood extent (section 5.4.1). For the complete Dhaka metropolitan area, the urban flood extent increased by a factor 6.4 from about 35.8km$^2$ in 2005 to 229.4 km$^2$ in 2050. This is close to the estimated 6.8 times damage increase between scenario 1 (baseline) and 2 (2050 urban growth).

Although this seems to add to the robustness of the outcomes, the observation might be coincidental. Due to the differences in the geographic extent of the two food maps, as well as the inherent difference in the calculation between the urban flood extent and damages, a correspondence seems hardly a plausible proof of some correlation or causality. Yet, in both cases the outcomes show that Dhaka's urban unbridled development is indeed a cause of concern. This outcome should definitely serve as a signpost for the representative authorities to take action because even with a moderate flood hazard, the growing population and assets in flood-prone areas surge future flood risk level. Investments in both flood mitigation and 'smart urban planning' are therefore essential to maintain Bangladesh's ambition to become a middle-income country by 2021 (BOI, 2014).

## 7.7 Discussion

One of the basic assumptions for this study was that no additional flood protection measures are taken in the growth projection interval (i.e. 2005-2050). This includes plans for an eastern embankment along the Balue River as well as incremental elevations made on building plots. The construction of the embankment is still under debate due to the financial and legal obstacles since the space is partially occupied by structures. Furthermore, the construction very likely leads to increased drainage blockages, which makes a decision a trade-off between fluvial and pluvial flooding (Montz and Tobin, 2008). Elevation of plots is often based on flood levels reached in past events and propagates through the area as "local knowledge".

A range of factors and assumptions could lead to under- or overestimation of the produced flood damages. This includes the limited focus on direct damages to commercial and residential properties, thus omitting a multitude of other assets susceptible to flood impacts. Furthermore, indirect damages due to business interruption or ripple effects in supply chains were omitted. Especially because many major floods in Bangladesh last for a considerable amount of time, the resulting economic disruptions are often large and therefore represent a significant share of the total flood damages. On the other hand, various behavioural aspects might limit the expected damage levels. Dhaka has a long flood history and most residents have experienced a number of flood events first hand. The subsequent coping and recovery capacity should therefore be considerable, thus limiting the extent and severity of suffered flood damages.

# 8. Further explorations

Sections of this chapter are based on:

Pathirana, A., Denekew, H. B., Veerbeek, W., Zevenbergen, C., & Banda, A. T. (2014). Impact of urban growth-driven landuse change on microclimate and extreme precipitation—A sensitivity study. *Atmospheric Research*, 138, 59-72.

## 8.1 Urban growth modelling and implications on water supply and sanitation planning

### 8.1.1 Introduction

The UN-Millennium Development Goals Report, (2011) states that in 2008, about 1.1 billion people did not have access to sanitation facilities and practised open defecation. This poses enormous health risks especially to the poor, of whom the majority resides in often rapidly developing slums. Since many of these slums are located along streams, health risks are propagated downstream and create serious environmental impacts further. Often these are aggravated by poor solid waste management, which leads to increasing pollution loads in surface water and floodplains. Rapid urbanisation, which in some cases is sometimes disproportionate in slums due to rural-urban migration of a relatively poor segment of the population, could in time lead to environmental calamities as well as severe health impacts. This is especially prudent in sub-Saharan Africa, where the currently majority the population is still living in rural areas. With the exception of Lagos and possibly Kinshasa, no megacities currently exist in sub-Saharan Africa. Rural-urban population equilibrium is expected only in 2050 (UN Habitat, 2009).

A better perspective on the future consequences of the growth of slums on potential pollution loads in streams in extensive metropolitan areas that are already reaching the limits of their capacity to convey and absorb wastewater is crucial to ensure a sustainable future for many cities in the developing world. An important requirement for the development of such an outlook, is the fact that an approach is needed that is spatially explicit. After all, both expected size of the slums as well as their respective location along the stream network is essential for developing future projections. Potentially, the outcomes of such an assessment show the need for a fundamental review of current planning practises or to invest in major infrastructural development to facilitate better sanitation and waste management. Good management of the rapid urbanization process and future urban growth could be achieved through effective land use planning, resource mobilization and capacity building UN-Habitat, (2007).

The objectives of the study were thus to explore the possibilities of using urban growth models in slum development projections and to schematically assess how the spatial distribution of that growth affects pollution loads in an existing stream network. As a case study, the rapidly developing megacity of Lagos has been chosen. Due to its extensive slum areas located along a stream network which discharges into a lagoon which serves as an important ecological and economic function, this city seems to be an ideal testing ground for such an endeavour.

## 8.1.2 Lagos

With over 10 million inhabitants in 2010 (Brinkhoff, 2012), the Nigerian capital Lagos is one of the two megacities in Africa. Although the city is a major economic hub in West Africa, two-thirds of the population lives in slums (Morka, 2007). Many areas still have inadequate sanitation facilities (Wugo et al., 2003) and less than 1% of the households are linked to a closed sewer system (Gandy, (2006). The majority uses onsite sanitation facilities ranging from bush (no system), pit latrines, VIP (Ventilated Improved Pit) latrines, and flush toilets modified into pour toilets (due of lack of water). According to the World Bank (2006), with refuse and raw sewage being swept in, floods in the slums are on an average knee deep both inside and outside the houses, and are said to last over five hours, causing immense economic hardship, and are a severe health hazard due to poor drainage systems. In Figure 42 such conditions are illustrated for the Makoko slum in eastern Lagos. Wugo et al (2003) state that the principal environmental problems in Metropolitan Lagos are related to the collection, treatment and disposal of sewage and wastewater. Because of these shortcomings, a substantial portion of untreated household sewage ends up in the lagoon which serves a major function for fisheries. The water table in most parts of Lagos is high hence bringing the high risk of pollution to the groundwater.

**Figure 42:** Children in Makoko slum in Lagos. Source: NOVA Next

## 8.1.3 Methodology and outcomes

### 8.1.3.1 Urban Growth Model and projections

The urban growth model and resulting outcomes are similar to those described in chapter 3. For the case of Lagos, an important assumption is made that has important consequences for the interpretation of the future projections: the high density built-up areas are characterised as shanty towns or slums. While these were initially identified as a separate LULC class, their delineation proved to be complicated and open to inter-pretation. The spatial characteristics of designated slum areas in Lagos like for instance Makoko, hardly differ from adjacent high density urban areas. This becomes clear from Figure 43 where the Makoko slum dissolves into a 'regular' high density built-up area. Given the statistics derived from literature (see previous paragraph) with a majority of the population living in slum-like conditions, the decision was made to classify all high-density built-up areas as slums. Apart from this LULC class, two other "regular" urban LULC classes were derived form the classification: medium and low density built-up areas.

To estimate future pollution loads for slum areas, an approximation is required of the future population residing in the slum areas. To obtain this, the density per urban built-up LULC class was derived by regression of the built-up areas on the total Lagos population (UN, 2011). Thus, the estimated historic changes in Lagos population were

**Figure 43:** Google Earth ™ aerial photo of the Iwaya neighbourhood (left) transitioning into the lagoon oriented Makoko slum (right).

associated to the observed changes in area covered by low, medium and high-density built-up areas. This lead to estimated densities of about 41, 70 and 703 inhabitants/ha for low, medium and high density areas respectively for the base year 2010.

As for all other cities in this research, the projections for Lagos are developed from 2010 onwards with iterations of 5 years and a horizon set at 2060. The resulting LULC maps are presented in Figure 44.

What can be clearly perceived is that the urban extent of southern Lagos remains relatively static since it's constrained by the lagoon and sea. Also on parts of the western periphery, the 2010 contour remains relatively stable. Lagos mostly expands on northern and eastern direction and ultimately merges with the cities of Ikorodu and Shagamu (east of Lagos) as well as with Abeokuta (north of Lagos). The projected development towards Abeokuta mostly consist of ribbon development along A5 highway that connects the northern stretches of Lagos (i.e. Ota), through Ifo to Abeokuta. The existing slum areas covering the south-eastern part of Lagos expand gradually to the north and west to fill up the central-southern urban extent almost completely in 2060. More dispersed slum formation can be observed in the northern extent of Lagos beyond 2035.

These trends are supported by some of the base statistics derived from the LULC projections. Between 2000 and 2010, Lagos almost doubled in size (see Figure 45). This growth trend continues into the projected medium and long term, with associated growth rates of about 87% for the interval 2010-2035 and 47% for the interval 2035-2060. Within the 2010-2060 interval, the proportion of slums compared to the complete urban extent is expected to increase from about 14% to 32%. The contribution

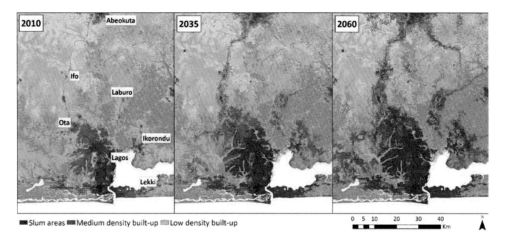

**Figure 44:** Observed land cover map of 2010 and simulated land cover maps for 2035 and 2060

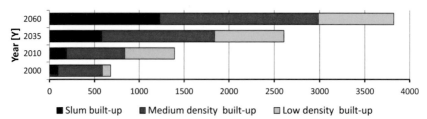

**Figure 45:** Urban growth statistics per urban LULC class for 2000, 2010 (observed), 2035 and 2060 (projected)

of planned areas (i.e. compounds for upper class citizens on the other hand decreases from 39% to 22% while ratio areas with medium intensity occupation remains static at about 48%.

## 8.1.4 Watershed Delineation and pollution loads

Watershed and stream network delineation was performed using a standard GIS-based approach on Aster GDEM data, which was also used as one of the thematic maps for the urban growth model of Lagos. By determining flow directions and subsequent flow accumulations the stream network was derived as well as the watersheds. The threshold value determining the minimum size of the watershed was set at 90 km2 (100,000 grid cells), which allows for relatively small basins.

The obtained watersheds were superimposed on the land use maps with high density built-up areas representing the major slum locations for the projection base year (2010) and for the medium (2035) and long term (2060) horizon. Using current population densities, the cumulative pollution loads were calculated using load factors derived by Henze and Comeau (2008) for the obtained stream network. These included reference values for the biochemical oxygen demand, chemical oxygen demand, Nitrogen, Phosphorus and total suspended solids of 60. 90, 5, 2 and 60 litre per capita and day respectively.

Note that this underlying assumption for this method is that the bulk of the wastewater produced in the slum areas drains into the stream network, which possibly creates an overestimation especially for watersheds saturated by slum areas. In other cases, wastewater stemming from non-slum areas might compensate for this overestimation. Since no actual data exist about these ratios, these generalisations have been applied.

The outcomes are presented in Figure 46, which shows the 2060 slum extent, the contours of the watersheds as well as the stream network. At each junction or outlet, a bar graph is placed which represents the estimated pollution loads for the years 2010, 2035 and 2060.

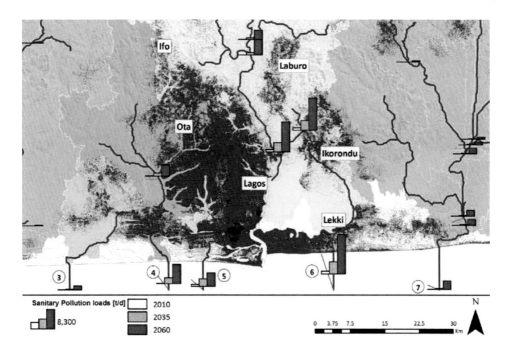

**Figure 46:** Sanitary Pollution Loads in slums of Lagos- 2010, 2035, 2060

The bar graphs in Figure 46 clearly illustrate the differences in pollution loads per junction of outlet, which is determined by the estimated slum population per watershed. These obviously show high levels for streams penetrating vast slum areas in combination with large watersheds. What is more important though are the estimated changes as a function of the projected population growth. The estimated increase in the interval 2010-2060 ranges from a factor 7 for outlet 5 to about 170 for outlet 3. Furthermore, in some cases (e.g. outlet 8, 10, 11) the initial loads were close to nil in 2010 but significantly increased in the projection period due to slum development in the watersheds. The calculated loads for the outlets presented in Figure 46, are presented in Table 16. The loads for the other outlets, can be found in Mudenda (2012).

**Table 16:** Estimated pollution loads

| Outlet # | Sanitary pollution loads [T/d] | | |
|---|---|---|---|
| | 2010 | 2035 | 2060 |
| 3 | 10 | 300 | 1700 |
| 4 | 200 | 2500 | 8200 |
| 5 | 800 | 3300 | 5800 |
| 6 | 1200 | 5100 | 16600 |
| 7 | 10 | 300 | 3500 |

### 8.1.5 Discussion

The results obtained from the analyses show that slums have the highest population densities with highly compacted housing arrangement. It is from this perspective that conversely these areas have the highest sanitary pollution loads as the areas do not have appropriate sanitation facilities.

It is important to acknowledge that provision of water supply and sanitation in itself can have a positive feedback effect on slum growth. Incorporating these feedbacks in future modelling exercises can could provide insight in alternative growth scenarios and associated spatial distribution of the estimated pollution loads. This in turn creates a foundation for alternative investment schemes and probably a different prioritisation. Since the outcomes suggest that slums are growing westwards, it would be wise to start planning well in advance to mitigate the associated stresses on the streams.

The model is based on a large number of assumptions and includes an abstraction level which renders the outcomes as merely indicative. Apart from various modelling issues, on of the main uncertainties are socioeconomic developments and subsequent investments in sanitation, behavioural changes of slum dwellers as well as more data-supported estimations about pollution loads in the existing stream network. Further study as well as expert judgement from local stakeholders is required to sustain these claims. The current project merely acts as a proof-of-concept and a novel application of urban growth modelling in long term environmental planning. As such, this exercise attempts to illustrate yet another water related area in which spatially explicit growth scenarios could be utilized.

Even so, the obtained outcomes can help authorities both in the planning office and policy makers to have a more concrete discussion about Lagos future environmental challenges. In turn, this might lead to more detailed research and applications of such an approach. Especially in the context of sub-Saharan Africa, where the process of extensive urban growth just starting, insight in the consequences and potential transitions such developments might produce is crucial.

## 8.2 Urban growth and microclimate

### 8.2.1 Introduction

The urban water cycle and the local climatic environment is invariably affected by the urban growth (Foley et al., 2005). Currently, there is an increasing body of evidence that the changes in the radiation and heat balance due to changes in surface albedo and vegetation cover can have significant impacts on the precipitation patterns in

(Watkins and Kolokotroni, 2012). These changes are to some extent caused by the urban heat island (UHI) (e.g. Sagan et al., 1979). There have been many empirical investigations suggesting that urban growth and the resulting UHI causes modulation of precipitation (e.g. Shepherd, 2006; Jauregui, 1996; Subbiah et al, 1990). Meir et al. (2013) examined two 2011 heat events in the New York City to evaluate the predictive ability of a Coupled Ocean/Atmosphere Mesoscale Prediction System (COAMPS) model. The model was able to capture the key features of the heat events, where urban rural temperature differences were as high as 4-50C.

Numerical modelling experiments are extremely relevant in understanding and quantifying the possible effect of UHI on rainfall. Research in this field only started to take off recently. For instance, Shem and Shepherd (2009) showed for increasing urban densities, a subsequent increase in UHI induced rainfall quantities between 10% and 13% for Atlanta, USA. Lin et al (2008) extended this approach by integrating for instance the influence of mountainous areas which affects rainfall on the leeward city of city. Yet, also opposing results were obtained by for instance Ntelekos et al. (2008) concluded that UHI did not contribute to a heavy rainfall in a controlled numerical experiment.

In the following sections, the results are presented of a series of numerical experiments conducted using a state of the art, 3D mesoscale atmospheric model – WRF-ARW (Skamarock et al. 2007) – in order to attempt to understand the impact of urbanisation-driven land use change on extreme rainfall events in and around cities. The model is applied to the rapidly growing metropolitan area of Mumbai, for which a spatially explicit urban growth scenario has been developed based on the extrapolation of past growth trends. Rainfall events observed in 2006 are compared to the urban growth adjusted events in 2060. The outcomes are analysed using standard statistical techniques used in rainfall frequency analysis to interpret the results in the context of urban storm drainage design and urban flooding

The presented outcomes are based on earlier work and a subsequent publication (Pathirana et al, 2014) in close collaboration with the author. The LULC change model and urban growth scenarios in this study were developed by Veerbeek et al (Djordjević et al, 2011; Veerbeek et al, 2013), while preliminary outcomes were presented at the International Conference of Urban Drainage (Veerbeek et al, 2011).

## 8.2.2 WRF-ARW Model

In order to assess the hydro-meteorological impacts of the projected urban growth, the meso-scale atmospheric model (WRF-ARW) was coupled with a land-use model with vegetation parameterization (Noah LSM). The WRF-ARW model numerically solves the four conservation relationships, namely mass, momentum and heat conservation of air and mass conservation including phase changes of water, by a non-hydro-static 3D set of equations. The model uses terrain-following vertical coordinate system and square grid horizontal coordinates with vector and scalar quantities staggered on the grid. The bottom boundary is provided by a surface scheme that in this case uses a LULC model that explicitly considers the vegetation and moisture effects of the surface. A full description of the WRF-ARW model is given by Skamarock et al. (2007). The model is suitable for both operational use (e.g. weather forecasting) and research studies.

## 8.2.3 Mumbai case-study with future urbanisation

To explore the future urban growth extent for the city of Mumbai, we used the model developed by Veerbeek et al (2015) was used to extrapolate past urban growth trends for future projections. The outcomes are shown in Figure 47. For the chosen areas, the urban extent of Mumbai and its suburbs increases in 2005 by about 22% (to 485 km2) compared to the base year 1990 (398 km2). The growth model estimates a less substantial for the midterm year of 2035 (Veerbeek et al, 2011) in which the urban extent further increases by about 13% (547 km2). Urbanisation mainly takes places in the eastern part of Navi Mumbai and the northern city of Thane. While currently disjoint, the outcomes predict the cities to merge with Mumbai which has little possibilities for expansion to the South/West because of its location on a peninsula.

The WRF model typically uses USGS LULC data, which covers the globe at a resolution of about 1 km. The LULC maps produced in this study are of a much higher resolution, and therefore had to be resampled and reclassified according to the USGS classification scheme in order to be used in the model. To accommodate for LULC changes, the albedo for built-up areas was reduced by 20% and the vegetation fraction by 75%. For this numerical experiment, three nested domains were constructed with resolutions of 30km, 6 km and 1.2km. The urban growth model covers an area of about 1700 km 2, which is about half of the area of the innermost domain. In this simulation the

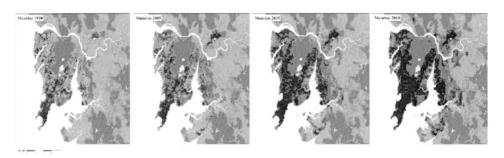

**Figure 47:** Urban growth of Mumbai metropolitan area 1990, 2005 (observed) and 2035, 2060 (projected)

WRF model was combined with the Noah land surface model to represent the surface processes, representing the interaction between surface and atmosphere (e.g. evapotranspiration).

Based on the simulation experiments for the four events, the aim was to analyse the possible change in the extreme precipitation frequencies using an intensity-duration-frequency (IDF) formula for western India proposed by Kothyari (1992). Kelkar (2005) estimates the two year return period rainfall in Mumbai as 200 mm. The outcomes were first analysed using a quantile-quantile analysis of all for events after which each of the 'Future' and 'Present' quantiles were assigned a return period based on the IDF curves.

### 8.2.4 Outcomes

Figure 48 (left) shows the distribution of rainfall intensities for quintiles at a 15 min temporal resolution, for the four historical events. The figure shows a clear impact of urban growth on to estimated rainfall. For three out of four events, the future rainfall is higher for moderate intensities (above 30mm/15 min) while for one event (event 4) the effect is limited to peak levels exceeding about 60 mm per 15 min minutes. In fact, event 4 shows a decrease in expected rainfall due to the project LULC changes which is another example the assumed relation is not valid in all conditions.

Figure 48 (right) shows the resulting shift of return periods under the projected future conditions of 2060. The analysis showed that the current 10-year rainfall event (75mm/h) would increase its frequency to a 3 year recurrence and 50-year (105 mm

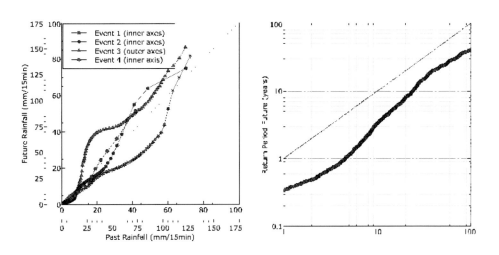

**Figure 48:** Quintile-quintile plots of rainfall intensities (left) and estimated present and future rainfall frequencies (right)

/h) to 22-year. These estimated shifts are significant for both frequent as well as extreme events.

The estimated changes in the spatial distribution for one of the rainfall events are illustrated in Figure 49, where the accumulated rainfall is shown for the present and future conditions. The figure shows a substantial increase in accumulated rainfall in the centre band of the image (between 19°00′ and 19°15′ latitude). In the growth scenarios, these regions show significant levels of densification due to expansion and infill. This includes the rapidly growing Navi Mumbai on the mainland, east of the Mumbai peninsula. In the northern part, covering for instance the city of Thane, the accumulated is somewhat less for the future projection which shows that the estimated changes in the spatial distribution of the rainfall are far from uniform. The dynamics of the urban heat island formation and the resulting changes in the rainfall are complex and depend on multitude of localized parameters like topography, surrounding land-use, features of the local/seasonal climatic regime etc. In order to make more reliable predictions on the impact each locality should be studied in detail. It is difficult to generalize the results at the global or even at regional level. This is an area that deserves further attention of the climate research community.

## 8.2.5 Discussion

The outcomes shown for the drainage impaired city of Mumbai are alarming. While

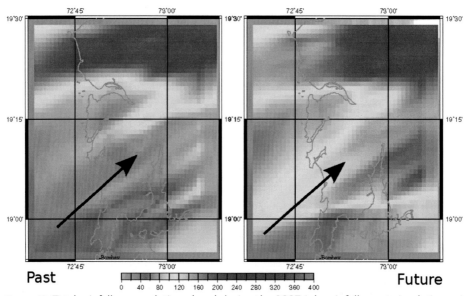

**Figure 49:** Total rainfall accumulations (mm) during the 2007 July rainfall event simulation using the 2005 (left) and projected 2060 (right) LULC map. The prevailing surface wind direction is marked by the arrow.

monsoon driven urban flooding occurs on an annual basis and extreme events cause severe impacts, precipitation levels are expected to increase if the current urbanisation trends continue. Latent heat is built up and increases the expected UHI when green areas diminish over time due to expansion of contiguous urban areas. In a way the increase of local precipitation completes the impact of urban growth on the flood risk cycle by also affecting the hazard component. So, apart from an increased exposure and possible vulnerability due to extensive growth, including extensive slum areas, the local rainfall patterns also intensify.

This conclusion urges for planned growth, in which extensive vegetation is integrated in the 'urban fabric', i.e. the spatial composition of new neighbourhoods. The question remains how representative the outcomes are for other cities. Currently, the question if similar results can be obtained for the other case studies is yet still unclear.

The model, input data and outcomes are based on a number of assumptions and simplifications. Apart from those affecting the outcomes of the urban growth scenarios, which have been discussed in earlier chapters, several issues affect the produced results. For instance, the level of enhancement of rainfall due to urban heat island in reality could be somewhat overestimated by these experiments. However, the outcomes produce strong evidence for proving the hypothesis that extensive urbanisation can cause changes in rainfall in and around cities. Hopefully, the outcomes have implications for urban planning, as well for the design of new urban drainage infrastructure that is expected to last for at least several decades.

From a conceptual point of view, the setup of the research is similar to that of the flood impact estimation for Dhaka, which is presented in the previous chapter. Also here, the effects a historic event (i.e. four rainfall events) are estimated for future conditions using a spatially explicit urban growth scenario. This shows the flexibility of this approach and the application of the growth model in different domains.

# 9. Towards an argument

## 9.1 Answering the RQs and testing hypotheses

In this final chapter an attempt is made to answer the research question that were put forward in the second chapter of this study. While these can be found throughout the respective chapters in this study, explicitly answering them in a separate section both provides convergence to the broad spectrum of topics that has been covered, and is essential for developing an actual argument and subsequent set of conclusions. Consequently, each research question is treated separately and complemented with the associated hypotheses testing.

> *RQ1. How can a context specific, spatially explicit urban growth scenario be developed that is based on a robust extrapolation of past trends, comparable data sources and with a relatively high level of detail and precision?*

The procedure for developing a scenario is described in Chapter 3, in which a spatially explicit LULC change model is introduced. The first part of the RQ introduces the question of local specificity, which is achieved by an extensive calibration phase in which an extensive set of parameters and LULC transitions rules of an initially generic model are optimised in order the model can mimic observed past LULC transitions. This calibration is highly localised, i.e. the local data (e.g. LULC base maps) used for the parameterization alter the model's behaviour significantly. This is captured in Hypothesis 1, where $H_1$ states that *"Two identical LULC change-models with different parameterizations are able to produce different LULC patterns from a single LULC distributions used as basemap data."* Although somewhat self evident, the setup of model calibration proves $H_1$. Every candidate solution in the MA-enhanced calibration sequence, consists of an alternative LULC distribution due to a different parameterization. A model calibrated for a different case study, with a subsequently different parameterization and transition rules, should therefore produce different growth projections for the same input data. Note that this does not necessarily mean that $H_0$ can be rejected. Theoretically, different models could produce identical LULC projections. In practise though, this is highly unlikely.

Robustness is first of all interpreted as a relative independency of case-study dependent local data, which is not available for other case study areas. In other words, the model only requires publically available data with a global coverage. This ensures that setup, operation and outcomes are uniform for any given urban area in the world. This is achieved by focussing on the exploitation of 'hidden features' within datasets (e.g. cluster size distribution, fractal dimension, etc.) and use those to describe LULC transitions. This is a departure from information greedy-models that rely on extensive datasets. Robustness furthermore requires consistent projections for multiple model runs. Given the probabilistic nature of the model, this is not necessarily ensured. The LULC transitions depend on a suitability-based rank order in which each cell's probability is described for a transition between two LULC-classes. Given the sheer amount of cell transitions in a given iterations, the stochastic assignment of cell transitions has only a limited effect; differences in produced LULC patterns can only be observed at the very local scale (i.e. clusters of 10s of cells). Since all case studies cover megacities, with associated LULC maps of 10s of millions of cells this influence is neglectable and the robustness of consecutive projections is ensured. If the model would be applied to small villages, or to cities where growth is very limited, consistency might be compromised.

The 'extrapolation of past trends', which is an essential feature in developing a BAU scenario for urban growth, touches upon the ratio between the epochs covered by the time series data and the projection periods. With a few exceptions, the growth models for the respective cities have been calibrated using data that covers a period of about 20 years (1990-2010). Yet, the horizon is set at 50 years (i.e. 2060). This suggests an imbalance that compromises the reliability of the projections. Yet, throughout this study the scenario is deliberately referred to as a projection instead of a prediction. This implies that no likelihood can be assigned to any of projections. Instead they are simply describing a spatially explicit multiplication of an observed spatial trend for a given period. This argument might seem merely semantic, but is essential in how the projections should be treated.

The issue of detail can only be answered from a comparative perspective; when compared to the applied flood maps describing riverine flooding, the level of detail is high (30m vs 30 arcsec). The model's accuracy is evaluated by assessing the so-called minimum mean similarity while ensuring a consistent value for the FD. Additionally, for each model a confusion matrix is constructed and the associated Kappa value is calculated (see 3.3.2 and Table 4). Application of these evaluation criteria is common practise in the domain of LULC-modelling. This issue coincides with Hypothesis 2, where $H_0$ can be rejected and $H_1$: *"The developed LULC change-model is able to produce LULC distributions that correspond to observed LULC distributions using evaluation criteria generally accepted in the current state-of-the-art in LULC-change modelling"*, holds.

**RQ2**. *How can urban growth affected flood risk be estimated and compared given the considerable size as well as the differentiation within and across the metropolitan areas?*

While the concept of flood risk is well defined (e.g. Zevenbergen et al, 2011), application is in most cases limited to a particular domain or region. The outcomes as illustrated in for instance in Section 7.5.2, illustrate direct flood damages but do not include indirect or intangible flood impacts. This is simply due to the extensive data requirements needed to develop a proper assessment. In this study, the flood risk assessment has mainly limited to the exposure of built-up areas to floods and to the drainage characteristics of the resulting built-up areas. In a sense, this consistently fits with the method in which the topic of urban growth has been operationalised, with a strong focus on the spatial propagation of built-up areas. The growth projections did not explicitly include population or economic attributes, but merely focussed on the intersection of urban and flood extent. Furthermore, for the assessment of riverine flooding, a more comprehensive appraisal of flood risk would not fit to the level-of-detail provided of the GLOFRIS flood dataset (i.e. 30 Arcsecs) and to the limited expressiveness of the LULC classification. The assessment therefore primarily focussed on (i) the development of the urban flood extent as a function of urban growth, (ii) flood depth distribution in the urban flood extent and (iii) the proportionality of the projected urban growth within and outside the flood extent. For pluvial flooding, the assessment was confined to the assessment of three proxy indicators that attempt to explain the drainage characteristics at macro-, meso- and microscale level. At macroscale, the focus was on estimating the average ISR within the urban footprint. At mesoscale on the FD and the open land ratio, which identify the level of fragmentation and perforation of patches covered by built-up areas. At microscale finally, the focus is on the transition within different classes of built-up areas. Do high density urban areas, with high ISRs, develop due to the transition from suburban areas or do they directly develop from urban extension or leapfrogging development. Furthermore, the indicator on micro-scale level also expresses an important characterisation of the development of future flood risk: the proportionality.

In this assessment, the issue of proportionality future flood risk development is an essential concept. While the different cities differ substantially in many dimensions, the comparison is focussing on the way the cities deviate from projected past trends. For instance, the impacts an identical flood in Beijing and Dhaka differ dramatically due to the different composition built-up areas. This prevents meaningful outcomes when directly comparing the two. Yet, when the comparison is based on the growth affected relative increase, the comparison is focussing on the rates at which the future flood risk is changing. This normalisation has been applied for both riverine and pluvial flooding (section 5.4.1 and 6.2.2.3).

***RQ3***. *What are the spatial characteristics of the growth projections and subsequent flood risk for the selected areas?*

One of the main outcomes of this study is that clear spatial growth trends within and between the cities, cannot be identified. The conclusions in Chapter 4, state that: "Urban growth is differentiated and not easily captured in a single driver or predictable growth pattern". The local conditions, conceptualised as growth constraints, push- and pull factors, differ too much to result in similar spatial characteristics. The outcomes from Chapter 4 extend to Chapter 5 and 6 where the projections are used to evaluate the changing future flood exposure. For riverine flooding, an initial characterisation was developed by analysing how the urban flood extent progresses over time as a function of urban growth (see Figure 22). Apart from the fact that the extent increases over time, a few particular features have been identified that separate some cities from others. For Beijing, Manila and Lahore, the extent seems to increase exponentially. Yet, other features (e.g. exponential decay) are limited to individual cities. The occurrence of exponential increase and decay of the urban flood extent proves $H_1$ of Hypothesis 3: *"Different spatiotemporal urban growth characteristics within the same flood extent of a given metropolitan area, will result in different spatiotemporal flood risk characteristics"*. In the assessment, the actual flood conditions (i.e. the set of inundation maps) remain constant during the 2010-2060 interval. The changes in the urban flood extent are therefore resulting exclusively from the changes in built-up areas.

For pluvial flooding, the differentiation in outcomes seems even more explicit. Analysis of the development of the estimated avg. ISR (macro-level assessment) or the FD (meso level assessment) hardly provides clear similarities between cities. Only when assessing the open land ratios (Figure 33) similarities in behaviour can be found. For instance, for Beijing, Dhaka and Lagos the fraction of open land, which is important for drainage, remains relatively stable as a function of urban growth.

> **RQ4**. *In which cities does normalised flood risk increases at a disproportionate rate and do those cities confirm existing expectations?*

The term "disproportionate rate" in RQ4, focusses on a discontinuation in the identified trends in urban flood risk development. Obviously, this issue is sensitive to the chosen interval on which the qualification is based. For the 2010-2060 interval, disproportionate growth of the urban flood extent caused by riverine flooding is well illustrated in Figure 26 where the relative growth is shown within and outside the estimated flood extent. Except for Lagos, Mexico City, Seoul and Shanghai, the urban flood extent grows at a disproportionate rate for all cities in this study. Dhaka, Lahore and Ho Chi Minh City are characterised as extreme cases, showing relatively rapid growth of urban areas within the floodplains. For smaller intervals within the 2010-2060 range, results may vary within and between cities since the increase of urban flood extent is not linear process. With somewhat relaxed constraints, these outcomes confirm $H_1$ of Hypothesis 4 which states: *"The normalised flood risk of two cities with similar growth rates, can lead to a significantly higher normalised future flood risk in one city over the other"*. One condition that somewhat weakens this claim are that for no pair of cities, the growth rates are exactly identical. For instance the estimated doubling periods for Dhaka and Ho Chi Minh City still differ (35.7 versus 35.9 years respectively). Furthermore, the concept of flood risk has been loosened throughout this study. For riverine flooding it was limited to the urban flood extent. For pluvial flooding, the question can be answered from the results presented in chapter 6, in particular by the analysis presented in Figure 34, illustrating the generalised proportionality of built-up areas, which associates to high ISRs. Here, the projected increase is set against a mere statistical extrapolation. All cities show a disproportionate increase. This holds particularly for Guangzhou-Shenzhen, Mexico City, Manila and Tehran with a relative increase of 25% or higher. On the opposite side of the spectrum lays Jakarta, where the projected increase generalised increase of less than 5%, which is only marginally disproportionate.

The issue if the cities for which the urban flood extent increase at a dispropor-tionate rate are the same cities that are often identified as those most suscepti-ble to future flood impacts, is somewhat more difficult to answer. One possible reference is to the outcomes and ranking developed by Nicholls et al (2008), fo-cussing on coastal flooding. Especially the ranking received considerable atten-tion in the media and is often used to illustrate which cities are most vulnerable to future flooding. Since the set of case study cities differs in this study differs, the obvious answer to this question would be "no". Some of the cities in this as-sessment are simply not located in coastal areas and are therefore not included in the ranking for future coastal flooding (e.g. Beijing, Mexico City, Tehran). Yet, a somewhat more insightful approach would be to compare the rank of the cit-ies that are present in the coastal ranking. The outcomes are presented in Table 17, where the ranking from Nicholls (ibid) is based on estimated exposed pop-ulation in 2070. For riverine flooding the ranking is based on the estimated size of the urban flood extent (see Section 5.4.1) and for pluvial flooding on the avg. ISR (see: Section 6.2.2.1). For most other cities, the differences in ranking are considerable. Only Dhaka is present in the top 10 of all 3 rankings. This not only reinforces the belief that the city is exceptionally flood prone, but also extends that notion to different types of flooding. Many of the other cities, show a high ranking for coastal flooding and for either riverine or pluvial flooding. This is the case for instance for Guangzhou-Shenzhen, which tops the riverine ranking and is located in 4th position for coastal exposure. Apart from Dhaka, Mumbai scores high in pluvial flooding (6th) while located in 2nd position for coastal flooding. Apart from these two examples, many others can be found. These outcomes are tested in Hypothesis 3, where $H_0$: can be rejected: *"The resulting rank list of cities based on future flood risk shows the same order than produced in current and past studies (e.g. Nicholls et al, 2008)"*. What is more important though is that the presented outcomes again illustrate the significant differences between the exposure to different hazard types (coastal, riverine and pluvial). Obviously, this is to some extent due to different assumptions and assessment methods. Yet, unsubstantiated claims that simple extend flood risk from one hard type to another are disproved.

Together these outcomes cover all the research questions and proposed hypotheses from chapter 2. They provide the foundations for the upcoming set of conclusions and recommendations.

## 9.2 Conclusions

In the assessment of future urban flood risk, the consequences of urban development are typically omitted. This is to some extent due to the perceived complexity of how cities will growth over longer periods of time. Yet, many areas around the world have witnessed unprecedented urban growth which lead to the appearance of megacities hosting populations of more than 10 million inhabitants. Especially in Asia, many of those cities are still showing very high growth rates which will significantly change the order of the world's largest urban agglomerations in the coming decades. Many of these areas are located within deltas or further upstream along major rivers, making them increasingly susceptible to future flood impacts. Furthermore, many are located in monsoon driven climates and are exposed to torrential rainfall way above the design standard of the stormwater drainage system. In comparison to climate change, the consequences of urban growth on future flood risk are considerable if not larger. That does not mean that the consequences of urban growth on flood risk can be assessed by mere statistical extrapolations. Local growth characteristics cause a large variability within and between cities' expected future flood risk profile. A quantification and qualification of these differences is the prime objective of this study, where future flood risk estimations are made based on spatially explicit urban growth scenarios for 18 rapidly growing megacities. The main focus point of the study is to differentiate between cities where flood risk might grow at a disproportionate rate and those where this is not the case. In other words, are there cities where growth trends lead to a relative increase of flood risk due to increased urban development in flood prone areas?

### 9.2.1 Riverine flooding

By combining detailed scenarios expressing land use and land cover changes with inundation maps, associated to flood scenarios with a range of different return periods, insight is provided in how flood exposure and depth distribution changes until 2060 with 5 year increments. The outcomes include a comparative assessment on how the future flood extent changes over time as well as a future ranking. Furthermore, the projected urban development is compared within and outside the estimated flood contours. This provides insight into the proportionality of the projected urban growth in relation to the current distribution of the urban flood extent.

The outcomes show that under a business-as-usual scenario, the majority of cities will face a considerable growth of the urban flood extent. In especially the Guangzhou-Shenzhen metropolitan area, but also in Calcutta, Beijing, Shanghai and Delhi the urban flood grows to areas covering several hundreds of square kilometres. Especially

**Table 17:** Ranking of cities exposed to coastal, riverine and pluvial flooding

| | Coastal | Riverine | Pluvial |
|---|---|---|---|
| Rank | 2070 | 2060 | 2060 |
| 1 | Calcutta | Guangzhou-Shenzhen | Karachi |
| 2 | Mumbai | Shanghai | Dhaka |
| 3 | Dhaka | Calcutta | Beijing |
| 4 | Guangzhou-Shenzhen | Beijing | Istanbul |
| 5 | Ho Chi Minh City | Delhi | Mexico City |
| 6 | Shanghai | Dhaka | Mumbai |
| 7 | Bangkok | Jakarta | Manila |
| 8 | Yangon | Ho Chi Minh City | Tehran |
| 9 | Miami | Mexico City | Seoul |
| 10 | Hai Phong | Lagos | Lagos |
| 11 | Alexandria | Manila | Ho Chi Minh City |
| 12 | Tianjin | Cairo | Cairo |
| 13 | Khulna | Seoul | Delhi |
| 14 | Ningbo | Mumbai | Guangzhou-Shenzhen |
| 15 | Lagos | Lahore | Jakarta |
| 16 | Abidjan | Karachi | Lahore |
| 17 | New York | Tehran | Calcutta |
| 18 | Chittagong | Istanbul | Shanghai |

the figures for the Guangzhou-Shenzhen region are alarming: the current (2015) urban flood extent which is estimated at 668 km2, is expected to more than double in size to 1417 km2 in 2060. Also for Calcutta the extend more than doubles (198 km2 in 2015 and 409 km2 in 2060). Cities like Lahore, Karachi, Tehran and Istanbul are located on the bottom of the ranking, with areas covering only a few square kilometres. While the top and bottom regions of the rankings remain relatively stable, the flood extent in some of cities in the middle sections growth especially fast. This holds for Dhaka and Ho Chi Minh City that move from rank 9 to 6 and 12 to 8 respectively. The distributions of the associated flood extent, which serve as an indicator for the magnitude of potential flood impacts, differ widely within and across cities. For some of the cities, the estimated flood depths are relatively low. This is especially the case in Shanghai and Beijing, while in Mumbai the depths even for 10-year event often exceed 2 meters (for only a moderately sized flood extent). For most of the other cities, the estimated depths differ significantly between events and across distributions urban areas, thus prohibiting the identification clear trends across cities. Yet the main outcomes of this study focus on the relative changes in future flood risk as a function of urban growth.

When assessing the proportionality of projected urban development within and outside the estimated flood contours, a major portion of the cities show increasing growth rates within flood prone areas. This is especially the case for Lahore, Ho Chi Minh City and Dhaka. In fact, in 2060 more than 50% of Dhaka's built-up area could be located in flood prone areas. In Ho Chi Minh City this could comprise of 24% in 2060. The projected proportion of urban development within and outside flood prone areas is illustrated in the following figure. Cities located furthest away from the red line show disproportionate urban development, where those located below the dotted red line have shown a tendency to urbanize especially within the floodplains while those above the dotted red line develop mostly on higher grounds.

## 9.2.2 Pluvial flooding

Urban growth has negative consequences for the drainage capacity of cities: the development of built-up areas leads to extensive soil sealing which in turn often results in increasing surface runoff and subsequent flooding of local depressions. This is especially the case in many of the fast growing megacities in this study, of which the majority is located in monsoon driven climate regions characterised by torrential rainfall in the wet season. In combination with an often underdeveloped drainage system, inadequate monitoring and maintenance as well as poor solid waste management, existing stormwater drainage systems in those cities are increasingly overcharged in especially the older downtown sections. Furthermore, existing surface water and green areas that could serve as peak storages are disappearing rapidly due to an increasing urbanisation pressure. This leads to contiguous built-up areas with uninterrupted impervious surfaces.

Assessment of the current as well as future drainage performance and subsequent flood hazard at the complete city scale is not a straightforward procedure. Current 1D/2D coupled hydraulic models require extensive data, calibration and computational resources to produce overland flow maps. At the scale of a megacity this is in most cases not feasible. Furthermore, spatially explicit growth projections do not provide a level-of-detail that includes future drainage structures. This makes a traditional modelling-based approach difficult. As an alternative, often a proxy indicator is used based on the impervious surface ratio (ISR) and the associated runoff coefficient. Such an indicator is often calculated at catchment or regional scale.

In this study a set of indicators has been chosen that are adjusted to the scale level of the produced spatially explicit urban growth scenarios. The drainage indicators focus on a macroscale level of the complete urban footprint, the meso- and microscale level. At macroscale, they focus on the estimated mean ISR by associating a set of LULC classes to specific ISR ranges. At mesoscale level, the indicators focus on the fractal dimension and the open land fracture. These provide quantitative information about the fragmentation level of patches of built-up areas as well as of the fraction of open land in contiguous clusters of built-up areas. Higher levels of both indicators explain for a high level of penetration of blue-green infrastructure which can serve as local water storage. At microscale level, the assessment is focussing mainly on the proportionality of the urban development and particularly at the origins of LULC patches that transition into high density built-up areas. By comparing the rates to mere statistical extrapolations, an assessment can be made if for instance is showing massive densification of existing suburbs or if new high density urban clusters appear. This characterisation explains something about the drainage characteristics at patch level.

In the macro level assessment, cities that are currently performing worst are Karachi and Mumbai with an avg. ISR of 0.62. Cities that perform relatively well are Calcutta and Shanghai with an estimated avg. ISR of 0.42 and 0.41 respectively. After application of the urban growth scenarios the estimated ISRs for 2060 increase for all cities, albeit at different levels. Karachi is still performing worst, with an estimate ISR of 07.2 closely followed by Dhaka where the avg. ISR grew by almost 25% to 0.71. Calcutta and Shanghai are still at the bottom of the list with avg ISRs of 0.48 and 0.45 respectively. At a mesoscale level especially Cairo and Beijing are performing relatively poor. Both show a consistently low fractal dimension indicating a compact composition of the urbanised areas. Cities characterised by a high level of fragmentation like Calcutta and Manilla show high fractal dimension and perform relatively well for this criterion. Cities with low open land fractions are Dhaka, Tehran and Mexico City, where especially the latter two show a substantial decline over the projection period 2010-2060. At Micro-scale level Guangzhou-Shenzhen, Mexico City Manila and Tehran are performing poorly. The transition into built-up areas is disproportionately high at the cost of mainly suburban areas. In other words, the fraction of high density built-up areas is going up at the expenses of suburban areas. Thus, at a local scale drainage is becoming more challenging because the footprint of soil sealing structures at individual patch level is becoming larger. The outcomes have been summarized in a comparative scoring matrix, based on the perceived relative changes in the consecutive multiscale drainage indicators. Overall, significant relative decrease in drainage performance for the period 2010-2060 can be expected for Delhi, Dhaka, Guangzhou-Shenzhen, Istanbul, Karachi, Manila, Mexico City and Tehran. Compared to today, the combined decline of the assessed drainage indicators possibly leads to a significant increase in pluvial flooding if no significant investments in the respective drainage structure are made.

Obviously, at all three scale levels (i.e. macro-, meso- and microscale) urban planning can have a significant impact on maintaining or even increasing the urban drainage capacity as well as to make these cities more resilient against the effects of peak rainfall events. Measures include limiting the development of contiguous built-up areas at macro-scale level, ensuring ample open spaces (e.g. parks, ponds) at neighbourhood level to provide peak storage and buildings with limited footprints, pervious parking at micro scale. Also here, cities where a significant decline drainage performance is estimated might benefit more from such interventions than cities where the performance is already at a critical level.

### 9.2.3 Consequences for flood risk management

The outcomes of this study provide a clue for future flood risk management strategies. Cities where extrapolations of past growth trends lead to a disproportionate urban development in flood prone areas might benefit from a more managed urban growth strategy (e.g. by using and enforcing flood averse or flood sensitive zoning plans). Cities that grow in 'safe' areas but that still have to cope with a significant future urban flood extent might benefit more from flood hazard mitigation areas (e.g. flood retention or flood protections structures).

Finally though, outcomes from this study also complement some earlier large scale assessments that were limited to the interaction between urban development and the effects of future sea level rise. Although this study confirms some of identified risk for some of the cities (e.g. Calcutta, Guangzhou-Shenzhen), it complements the ranking with cities like Beijing or Delhi that might have been underexposed when it comes to increasing future flood risk.

## 9.3 Recommendations

Many detailed recommendations are already suggested in respective chapters. The following section merely provides a summary of those recommendations as well as a sporadic addition.

### 9.3.1 Urban growth model

Although the urban growth model and particularly the automated calibration method used to develop the growth scenarios in this study (Veerbeek et al, 2015) delivered robust and high quality projections, there are many aspects which can be further refined. These might range from alternative representations of the cell lattice, nested multi-scale models to asynchronous transition updates (Benenson et al, 2004; Veerbeek et al, 2012). In the chapter 2, where the model is introduced, a number of specific improvements have been discussed which particularly focus on the memetic algorithm and integration of the "hillclimber" to effectively reach local optima in the model's parameterization. The impact of these refinements is very uncertain. This is partly due to the lack of a systematic research in CA-based LULC-change models, which is confirmed by Grinblat et al (2016) who urges for a standardised dataset and testing procedure to assess a model's performance. An argument against such a procedure is that the currently available datasets associated with the case study as well as the actual purpose of the model dictate the actual model design. In the case of the model used in this study for instance, the goal was to develop a model for long term projections that could be applied for any urban area in the world. This implies a series of constraints on for instance the use of base maps, thematic maps as well as the conceptual treatment of time in the model. Recommendations are therefore primarily focussed on further enhancing performance while maintaining the model's application:

- *Improved exploitation of 'hidden' features in basemaps.* Integration of additional spatial metrics as additional dynamic maps that can be used as additional weights in the transition rules;
- B*etter facilitation of planning interventions*. To ensure a uniform treatment of particular planning policies or measures, the model requires more flexibility to integrate location based development constraints or extensions (i.e. local push factors) that alter the suitability of individual cells for transitioning into built-up areas. Furthermore, policies that affect for instance the housing preferences could be implemented as parameters that affect the compactness of patches representing built-up areas, therefore mimicking more dense or fragmented urban development.
- *Flexibility*. In chapter 8, the model is applied in the context of water quality and UHI-driven local precipitation changes. To further extend applications, additional cell attributes might be required to express key-variables for the specific application;
- Urban renewal. The current model primarily focuses on conversion of non-urban into urban areas, i.e. processes that explain urban growth. Yet, the issue of urban renewal, which represents a considerable fraction of the building activities in cities, is not represented in the model. This can be accommodated by including 'age' as an explicit cell-attribute as well as in the transition rules.

## 9.3.2 Riverine flood risk assessment

Also for the topic of riverine flood risk assessment, many recommendations have been already made (see chapter 3), including a more inclusive treatment of the concept of risk and the mismatch between the spatial resolution between the flood maps and urban growth scenarios. This is obviously due to the fact that flood maps produced in GLOFRIS, were originally not intended for use in context of individual cities. That does not mean that the model cannot be further refined; the current growth of data sources, improved sensors as well as crowd-based data sourcing make the production of high resolution flood maps feasible within the foreseeable future. The required computational power is provided by cloud-based clusters that were previously only available to dedicated institutes. An important quality improvement of the model would be better integration of manmade flood protection measures. Currently the flood extent is predominantly based on the elevations derived from DTMs, which might cause an overestimation of the estimated flood extent. Currently, efforts are made to complete the model with such features (Scussolini et al, 2015). If the GLOFRIS model would emerge as the standard in global flood modelling, resources to its refinement might be more focussed in the future. As a consequence, contribution from different international institutes (incl. Government institutes) can focus on improving the model instead of on the development of alternative models that provide a similar functionality.

## 9.3.3 Pluvial flood risk assessment

The assessment of pluvial flooding in this research was based on a set of proxy indicators which only provide a haphazard relationship to actual flood conditions. Ideally, a set of city-wide inundation maps associated to design rainfall events needs to be developed from which an urban flood extent and depth distribution can be derived. Yet, as already discussed in chapter 4, such an approach could require explicit integration of a stormwater drainage model which in case of an urban growth model, with no representation of future drainage structures, is unfeasible. Instead, a depression analysis could be made that indicates local depressions or even facilitates overland flow by redistributing rainfall quantities into a network of sinks. Since most of the cities in this research are located in relatively flat areas, on the alluvial plains of river deltas, a significantly higher resolution DTM needs to be applied to obtain expressive results. The currently resolutions provided by Aster GDEM or SRTM2-data which approximates 25m-30m resolutions is insufficient to represent local inundations of streets and open spaces. Yet, the rapid development in available areas covered by LIDAR-based ultra-high resolution elevation data should quickly make this approach attainable. Additionally, the required additional computational resources could be easily provided by cloud-based computing clusters. An important consideration is that a depression based approach, is particularly suited to indicated flood locations in case of extreme rainfall events, where the influence of the stormwater drainage network is limited. This is actually the case in most of the cities in this study, where monsoon driven extensive rainfall is far exceeds the limited capacity of an often outdated drainage system.

Finally, the Interaction of riverine and pluvial is in some cases a matter of concern. In many cities drainage outlets are not yet provided with non-return valves, which leads backflow into the drainage systems in case of a high river discharge. Such examples of drainage blockage are often responsible for a further boost in pluvial flood levels, which might need to be considered when developing flood scenarios. This extends also to coastal flooding where tidal influence can cause significant impact when coinciding with peak river discharges.

## 9.3.4 Additional flood hazards

Apart from the absence of storm surges from the assessment, integration of additional risk drivers might be required since they significantly amplify future flood risk levels. One of these factors is the on-going process of subsidence that lowers ground levels in many deltaic regions. Subsidence significantly boosts the susceptibility to pluvial flooding, prolongs flood durations as well causing an increase of flood depths for all types of floods. One of the main contributors to subsidence is groundwater extraction which in turn is amplified by the on-going process of urbanisation. It is therefore no surprise that subsidence occurs in many of the regions covered in this study including Ho Chi Minh City, Jakarta and Manila were annual subsidence rates reach levels up to 80mm, 100mm and 45mm respectively (Bucx et al, 2015). Obviously, climate change induced sea level rise will further increase the importance of subsidence in the overall flood risk profiles of delta cities.

Future studies should at least provide a method to include the estimated subsidence rates into the assessment. This might be in a spatially explicit fashion, where elevations are lowered using a similar approach in the downscaling procedure used to assess the sensitivity of the outcomes against the use of high resolution DTMs. Yet, inclusion of subsidence adds another level of complexity to the assessment, since ground water abstraction links directly to the urban growth rates used for the projections.

## 9.3.5 Scenarios

A possible recommendation would be to broaden the current growth scenario approach, by a growth rate-driven lower and upper bound. Indeed, a single projection for a future city might seem a somewhat simplistic approach for describing a possible state of a relatively complex system. As customary for IPPC-based climate change scenarios, an ensemble of scenarios is developed that account for a more moderate lower as well as an extreme upper bound of greenhouse gas emissions and subsequent impacts. In case of the developed urban growth scenarios, population-change adjusted growth rates could be applied that represent a lower, mean and upper urbanisation rate. Such an approach has been used for in the refined scenario for Dhaka, described in chapter 5. While simple in implementation, this raises the question about other parameters that determine the spatial distribution of future urban areas. For instance, clustering parameters or mean patch sizes could provide lower and upper bounds to better cover the bounded uncertainty that is intrinsic to the development of long term spatially explicit urban growth projections. In short, one could argue that this approach is "opening Pandora's box" and moves away from the primary objective of the model which is to develop robust extrapolations of observed spatial trends in urbanisation into the future. Obviously, this is a mere academic discussion and might move away from practical demands, a low, medium and high growth scenario are often common practise.

An alternative approach would be to cover the complete variable space and to approach scenario development as an exploratory modelling exercise. Evaluation of the outcomes could lead to identification of key parameters, which effectively would turn such an approach into a sensitivity analysis. Although interesting, the approach seems merely academic and provides insight in the robustness of the produced LULC distributions. Nevertheless, after development of formalised procedure, the approach can be used as an additional method for validation during the calibration stage of the growth model.

### 9.3.6 Assessment

An important extension of the presented assessment would obviously be the integration of climate change scenarios. This would provide further evidence about how the impact of future urban growth would compare to projected climate change induced changes in flood risk. Although a small exploration is included in Section 5.6, the outcomes only serve as a crude first approximation limited to the domain of riverine flooding. Unfortunately, during the period in which this study was performed, a downscaled version of the data which complies with the same specifications as the flood maps used for the assessment of riverine flooding was not yet available. Nevertheless, Integration of climate change scenarios should be a top priority in future development.

A further recommendation is to focus the assessment on the comparison of relative changes in projected urban flood risk. The differences in the composition of built-up areas, the associated flood vulnerabilities and ultimately the estimated impacts are too large to allow for a direct comparison between cities. Given the uncertainties that are associated to the spatiotemporal extent on which the analyses are made, even normalizing the impacts into monetary values neglects important differences between for instance Dhaka and Ho Chi Minh City. Nevertheless, additional research can be done to at least better express some of the density characteristics of cities (e.g. Berghauser et al, 2009) in order to ensure a more expressive use of the urban growth projections.

## 9.4 Discussion

The notion of a 'business-as-usual scenario' for urban growth requires a conceptual approach that is subject to some arbitrary choices. The statement in which the concept is explained as 'extrapolating observed growth trends into the future' opens up a wide range of interpretations. The Merriam-Webster dictionary states that business as usual refers to a state where "something is working or continuing in the normal or usual way", while the Cambridge describes it as "when things are continuing as they always do". This could imply a wide interpretation, in which many of the factors that determine urban growth are captured (e.g. trends in population growth, economic development or ground pricing) or a more constrained interpretation in which the focus is limited to a manifestation of urban growth: LULC changes. However, as already explained in the introduction this latter approach is taken due to conceptual (top-heavy model) as well as practical (limited data availability) considerations.

A question of a more conceptual nature is the issue if cities can actually reach a maximum size. While theoretically, the developed growth model could produce projections with a unlimited number of iterations, the produced scenario would produce an urban extent well beyond anything seen in the current conditions. Nevertheless, the current size of the Tokyo-Yokohama metropolitan area which hosts an estimated 37.8 million inhabitants (Demographia, 2015) would probably have seemed impossible. On the other hand, the cities of Dhaka and Lagos, which currently hosts around only 15.6 and 21 million inhabitants respectively (ibid) are found consistently at the bottom end of the liveability index (The Economist Intelligence Unit, 2016 ). A further growth of these cities might seem to lead to a collapse of basic services like infrastructure, utilities, healthcare, etc. Yet, even during the period of this study, both cities welcomed 100s of thousands of new inhabitants without any sign of a collapse. While at a global scale a number of scholars have demonstrated the limits of growth (e.g. Meadows et al, 1972; Meadows et al, 2004), such evidence has not yet been provided for the growth of cities. For instance in the Pearl River delta which seems to be transforming rapidly into the world's largest urban agglomeration, the limits of growth have yet to be found.

Ultimately, comparative studies like presented here will continue to develop in breadth and depth since data availability and computing power are increasing at an unprecedented pace. The current Landsat 8 remote sensing data for instance, widens the range for time series data on LULC (due to failure of the Landsat 7 sensor) and increase the accuracy of classification by adding 2 additional bands. LIDAR refined DTMs are setting the new standard in elevation models (e.g. Intermap, 2017) and cloud computing makes supercomputing available for a wide audience at competitive prices.

In some respects, reality already caught up with this study. The selection of rapidly growing megacities that was made in 2010, is currently outdated. Especially in China the number of megacities has more than doubled. In 2016, the cities of Chengdu, Chongqing, Harbin, Tianjin and Wuhan all reached the status of megacity (Demographia, 2016). In India, two additional megacities emerged: Bengaluru and Chennai. These developments provide a sharp contrast with a much more modest growth of megacities in the rest of the world. In all of Africa, only Congo's Kinshasa can added to the list. The estimates on which these classifications are based upon, are somewhat clouded since reliable population figures are often unavailable.

While the focus of this study is on fast growing megacities, the cities that are showing the fastest growth are small and medium-sized cities with populations of less than 1 and 5 million respectively (OECD, 2014). In China and India for instance, these cities are expected to absorb half of the urban expansion in the coming years. Compared to megacities, the transformations that occur in these types of cities undergo are often much more fundamental since they are often transitioning from local hubs dominated by a single economic activity to economies of scale. These changes often have a fundamental impact on the water system where significant investments in for instance the drainage system are required to keep up with the increasing pressures. Some insights in the potential future consequences of growth in medium-sized cities have been made in Can Tho, Vietnam (Huong et al, 2013). Yet, due to the volume of rapidly growing small and medium-sized cities a more systematic research is difficult. Possibly, studies should be performed on delta or catchment scale in order to include a larger number of these cities.

Beyond the research community, there are many uncertainties regarding the future demand of these kinds of studies. Large city network organisations like for instance the Rockefeller 100 Resilient Cities network or ICLEI provide support for individual cities but hardly invest in comparative studies. Comprehensive assessments often either focus on single issues (e.g. coastal flooding) or are limited to qualitative assessments of current conditions (e.g. UNISDR's New essentials for making cities resilient) . Wide support for prospective quantitative studies is limited. This was also the experience in this study, where progress was piecemeal and dependent on a partial overlap with projects and case studies; no single project covered the scope and breadth of the presented work. Nevertheless, one of the main contributions of this study is the focus on quantitative spatially explicit (i.e. GIS-oriented) outcomes. This makes the outcomes more 'tangible'' and accessible to a wider audience. The downside is that this also gives rise to an often biased discussion about the 'correctness' of the results; urban growth scenarios are often taken as predictions that sometimes do not confirm pre-existing ideas. This is obviously a merit of the study, but requires proper framing of the presented outcomes and conclusions, especially outside the academic world.

Finally though, further development and extensive investments in the adoption of a spatially explicit baseline for urban growth in future environmental assessments and policy development will likely define the agenda for the coming years. This requires dissemination, the tailoring of the outcomes to fit within actual projects and the renewed support of partner institutes.

# 10. Bibliography

Abidin, H. Z., Andreas, H., Gumilar, I., Fukuda, Y., Pohan, Y. E. & Deguchi, T. (2011, 06). Land subsidence of Jakarta (Indonesia) and its relation with urban development. Natural Hazards, 59(3), 1753-1771.

Acharya, S. S., Sen, S., Punia, M. & Reddy, S. (Eds.) (2016). Marginalization in Globalizing Delhi: Issues of Land, Livelihoods and Health. Springer

Adikari Y., Osti R. & Noro T. Flood-related disaster vulnerability: an impending crisis of megacities in Asia. J Flood Risk Manage 2010, 3, 185–191.

Agterberg, F.P. and Bonham-Carter, G.F. (1990). Deriving weights of evidence from geoscience contour maps for the prediction of discrete events. XXII Int. Symposium AP-COM, 381-395.

Ahmed, F., Gersonius, B., Veerbeek, W., Khan, M. S. A., & Wester, P. (2015). The role of extreme events in reaching adaptation tipping points: a case study of flood risk management in Dhaka, Bangladesh. Journal of Water and Climate Change, 6(4), 729-742.

Alam, M. S. and Hossain, M. S. (2012), [URL: http://en.banglapedia.org/index.php?-title=Beel], Banglapedia (Second ed.), Asiatic Society of Bangladesh, retrieved September 2017

Alam M. and Rabbani M.D.G. (2007) Vulnerabilities and responses to climate change for Dhaka. Environ Urban, 19, 81–97.

Allen, G. R. (1954). The 'courbe des populations:'a Further Analysis. Bulletin of the Oxford University Institute of Statistics, 16, 179-189.

Almeida, de, C. M., Batty, M., Monteiro, A. M. V., Camara, G. Soares-Filho, B. S. Cerqueira, G. C. & Pennachin, C. L. (2003) Stochastic cellular automata modeling of urban land use dynamics: empirical development and estimation. Computers, Environment and Urban Systems 27, pp481-509

Alonso, W. (1964). Location and land use: Toward a general theory of land rent. Cambridge, MA: Harvard University Press.

Angel, S. (2012). Atlas of urban expansion. Cambridge, MA: Lincoln Institute of Land Policy.

Angel, S, J. R. Parent & Civco, D. L. (2007) Urban Sprawl Metrics: An Analysis of Global Urban Expansion Using GIS. ASPRS May 2007 Annual Conference. Tampa, FL

Angel, S., Sheppard, S., Civco, D. L., & Buckley, R. (2005). The dynamics of global urban expansion. Washington, D.C.: World Bank, Transport and Urban Development Department.

Apel, H., Aronica, G. T., Kreibich, H., & Thieken, A. H. (2008). Flood risk analyses—how detailed do we need to be? Natural Hazards, 49(1), 79-98.

Arai, T., and Akiyama, T. (2004). Empirical analysis for estimating land use transition potential functions - case in the Tokyo metropolitan region. Computers, Environment and Urban Systems , 28:65-84.

Augustijn-Beckers, E., Flacke, J. & Retsios, B. (2011). Simulating informal settlement growth in Dar es Salaam, Tanzania: An agent-based housing model. Computers, Environment and Urban Systems, 35(2), 93-103.

Bäck, T., Dörnemann, H., Hammel, U., & Frankhauser, P. (1996). Modeling urban growth by cellular automata. Parallel Problem Solving from Nature — PPSN IV Lecture Notes in Computer Science, 635-645.

Bailey, T. and Gatrell, A. (1995) Interactive Spatial Data Analysis. Longman, Harlow.

Banerjee-Guha, S. (2002). Shifting cities: urban restructuring in Mumbai. Economic and Political Weekly, 121-128.

Battiti, R., Brunato, M. & Mascia, F. (2008) Reactive search and intelligent optimization. Springer, Heidelberg

Batty, M. (2007). Cities and complexity: Understanding cities with cellular automata, agent-based models, and fractals. Cambridge, MA: MIT.

Batty, M., & Longley, P. (1994). Fractal cities: A geometry of form and function. London: Academic Press.

Batty, M. and Xie, Y. (1994). From cells to cities. Environment and Planning B 21: 31–48.

Bari, M. F., and Hasan, M. (2001). Effect of urbanization on storm runoff characteristics of Dhaka City. In Proceedings of the congress-international association for hydraulic research, pp. 365-371

BBC (2016). Europe flooding: Five dead as waters rise in Germany and France - BBC News. [URL:http://www.bbc.com/news/world-europe-36429381] Retrieved September 2017

Bedient, P.B. and W.C. Huber (2002) Hydrology and Floodplain Analysis, 3nd ed., Prentice Hall, New Jersey, USA.

Beijing Statistics Bureau. (2005). Beijing statistics yearbook in 2005. Beijing: Chinese Statistics Press.

Beijing Municipal Planning Committee (2006) Beijing Institute of City Planning, Beijing Academy of Urban Planning. Beijing Urban Planning Atlas (1949-2005), 2006. (in Chinese)

Benenson, I., and Torrens, P. M. (2004). Geosimulation: Automata-based modeling of urban phenomena. Hoboken, NJ: John Wiley & Sons.

Benenson, I. and Omer, I. (2003). High-resolution census data: A simple way to make them useful. Data Science Journal, 2, 117-127.

Berghauser Pont, M. Y., & Haupt, P. A. (2009). Space, density and urban form (Doctoral dissertation, TU Delft, Delft University of Technology).

Berkhout, F., Hertin, J., & Jordan, A. (2002). Socio-economic futures in climate change impact assessment: Using scenarios as 'learning machines'. Global Environmental Change, 12(2), 83-95.

Berryman, A. A. (1992). The Orgins and Evolution of Predator-Prey Theory. Ecology, 73(5), 1530-1535.

Bhatta, B. (2009, 08). Analysis of urban growth pattern using remote sensing and GIS: A case study of Kolkata, India. International Journal of Remote Sensing, 30(18), 4733-4746.

Blanco, H., Alberti, M., Olshansky, R., Chang, S., Wheeler, S. M., Randolph, J., ... & Popper, F. J. (2009). Shaken, shrinking, hot, impoverished and informal: Emerging research agendas in planning. Progress in Planning, 72(4), 195-250.

Blecic, I., Cecchini, A. & Trunfio, G. A. (2014). Fast and Accurate Optimization of a GPU-accelerated CA Urban Model through Cooperative Coevolutionary Particle Swarms. Procedia Computer Science, 29, 1631-1643.

BOI (Board of Investment Bangladesh Prime Minister's Office). Message from Prime Minister's Office. 2014. URL[http://www.boi.gov.bd/index.php/welcome-message-2/primeminister-s-message] Accessed November, 2015

Bonham-Carter, G (1994). Geographic information systems for geoscientists: modeling with GIS. New York, Pergamon, p238-257

Brinkhoff, T. (2010) The Principal Agglomerations of the World. URL[https://www.citypopulation.de] Accessed September 2017

Brinkhoff, T (2012) : City Population, http://www.citypopulation.de Accessed: June 2012

Brown, R. R., Keath, N., & Wong, T. H. (2009). Urban water management in cities: Historical, current and future regimes. Water Science & Technology, 59(5), 847.

Brown, S., and Nicholls, R. (2015). Subsidence and human influences in mega deltas: The case of the Ganges–Brahmaputra–Meghna. Science of The Total Environment, 527-528, 362-374.

Brown R. R. (2005) Impediments to integrated urban stormwater management: the need for institutional reform. Environ. Manage. 36, 455–468.

Bruegmann, R. (2005). Sprawl: A compact history. Chicago: University of Chicago Press.

Bruijn, K. D. (2005). Resilience and flood risk management: A systems approach applied to lowland rivers. Delft: DUP Science.

Bruijn, K. M. (2004). Resilience indicators for flood risk management systems of lowland rivers. International Journal of River Basin Management, 2(3), 199-210.

Brun, S., & Band, L. (2000, 01). Simulating runoff behavior in an urbanizing watershed. Computers, Environment and Urban Systems, 24(1), 5-22.

Bruzzone, L and Serpico, S. B. (1997) An iterative technique for the detection of land-cover transitions in multitemporal remote-sensing images. IEEE Transactions on Geoscience and Remote Sensing, 35: 858–867

Bucx, T. H. M., van Ruiten, C. J. M., Erkens, G. & de Lange, G. (2015). An integrated assessment framework for land subsidence in delta cities. Proceedings of the International Association of Hydrological Sciences, 372, 485.

Caglioni, M., Pelizzoni, M., & Rabino, G. A. (2006). Urban Sprawl: A Case Study for Project Gigalopolis Using SLEUTH Model. Lecture Notes in Computer Science Cellular Automata, 436-445.

Caixin Online (2012) URL[http://english.caixin.com/2011-06-01/100265200.html] Accessed November 2016

Canters, F., Chormanski, J., Van de Voorde, T. & Batelaan, O. (2006). Effects of different methods for estimating impervious surface cover on runoff estimation at catchment level. In Proceedings of 7th International Symposium on Spatial Accuracy Assessment in Natural Resources and Environmental Sciences, Lisbon, Portugal 57, 557-566.

Cardona, O. D., van Aalst, M. K., Birkmann, J., Fordham, M., McGregor, G., & Mechler, R. (2012). Determinants of risk: exposure and vulnerability. Cambridge University Press, Cambridge, UK

Carlson, T. N., and Arthur, S. T. (2000). The impact of land use — land cover changes due to urbanization on surface microclimate and hydrology: A satellite perspective. Global and Planetary Change, 25(1-2), 49-65.

Chan, F. K. (2012). Flood risk appraisal and management in mega-cities: A case study of practice in the Pearl River Delta, China. Water Practice & Technology, 7(4), 1-8.

Chen, A. S., Hammond, M. J., Djordjević, S., Butler, D., Khan, D. M., & Veerbeek, W. (2016). From hazard to impact: Flood damage assessment tools for mega cities. Nat Hazards Natural Hazards, 82(2), 857-890

Chen, X. S., Ong, Y. S. & Lim, M. H. (2010). "Research Frontier: Memetic Computation - Past, Present & Future". IEEE Computational Intelligence Magazine 5 (2): 24-36.

ChinaSMACK (2011). URL[http://www.chinasmack.com/2011/pictures/torrential-rains-flood-beijing-subways- -and-forbidden-city.html] Accessed August 2017

Christaller, W., & Bashkin, C. W. (1966). Die Zentralen Orte in Süddeutschland. Central Places in Southern Germany; Translated by Carlisle W. Baskin. Prentice-Hall.

Clarke, K. C., Hoppen, S., & Gaydos, L. (1997). A self-modifying cellular automaton model of historical urbanization in the San Francisco Bay area. Environment and Planning B: Planning and Design, 24(2), 247-261

Clark Labs (2013) IDRISI Selva GIS and Image Processing Software URL[https://clark-labs.org/terrset/idrisi-image-processing] Accessed September 2017

Cohen, B. (2004). Urban Growth in Developing Countries: A Review of Current Trends and a Caution Regarding Existing Forecasts. World Development, 32(1), 23-51.

Comber, A. J., F., and Wadsworth, R. A. (2005). What is land cover?

Constanza, R. (1989). Model goodness of fit: a multiple resolution procedure. Ecological Modelling, 47, 199–215

De, U. S., Singh, G. P., & Rase, D. M. (2013). Urban flooding in recent decades in four mega cities of India. J. Ind. Geophys. Union, 17(2), 153-165.

Deltares (2015) Sinking cities, An integrated approach towards solutions. Deltares-Taskforce Subsidence URL[https://www.deltares.nl/app/uploads/2015/09/Sinking-cities.pdf] Accessed September 2017

Demographia (2010) Demographia World Urban Areas; Built-up Urban areas or World Agglomerations, 5th Annual Edition, April 2011. http://www.demographia.com/d-new.htm

Demographia (2014) Demographia World Urban Areas: 10th Edition (2014 Revision). 2014. URL[http://www .demographia.com/db-worldua.pdf] Accessed December 2016.

Demographia (2016) Demographia World Urban Areas: 12th Edition (2016 Revision). 2016. URL[http://www .demographia.com/db-worldua.pdf] Accessed August 2016.

Dewan A.M. and Corner R.J. (2014) Spatiotemporal analysis of urban growth, sprawl and structure. In: Dewan A. M. and Corner, R. J. (eds). Dhaka megacity. Netherlands: Springer, 99–121.

Dewan A.M. and Yamaguchi Y. (2009) Using remote sensing and GIS to detect and monitor land use and land cover change in Dhaka Metropolitan of Bangladesh during 1960–2005. Environ Monit Assess 150, 237–249.

DHI (2009) Technical Assistance for Beijing Water Environment Project. Technical Report, BEPII-BWC1

Djordjević, S (2010) Modelling pluvial flooding. In: Zevenbergen, C., Cashman, A., Evelpidou, N., Pasche, E., Garvin, S., & Ashley, R. (EDS). Urban flood management. CRC Press. Boca Raton, USA

Djordjević, S., Butler, D., Gourbesville, P., Mark, O., & Pasche, E. (2011). New policies to deal with climate change and other drivers impacting on resilience to flooding in urban areas: The CORFU approach. Environmental Science & Policy, 14(7), 864-873.

DNA India (2017) [URL: http://www.dnaindia.com/topic/coastal-road-project] Accessed September 2017

Donnelly TG, Chaping FS, Weiss SF (1964) A probabilistic model for residential growth. An Urban Studies Research Monograph. Chapel Hill, Institute for Research in Social Science, University of North Carolina: 65

ECTP. European construction technology platform (2005) Strategic Research Agenda for the European Construction Sector: Achieving a sustainable and competitive construction sector by 2030. URL [https://www.certh.gr/dat/8BB3421E/file.pdf] Accessed September 2017

Emberger, G. (2016). Urban Transport in Ho Chi Minh City, Vietnam. Sustainable Ho Chi Minh City: Climate Policies for Emerging Mega Cities, 175-191.

Engelen, G., White, R. & Uljee, I. 1997. Integrating constrained cellular automata models, GIS and decision support tools for urban and regional planning and policy making. In Decision support systems in urban planning, Ed. H. Timmermans, 125–55. London: E. & F. N. Spon Ltd.

Erdas Inc. (1999). Erdas Field Guide. Erdas Inc., Atlanta, Georgia  URL[http://web.pdx.edu/~emch/ip1/FieldGuide.pdf] Accessed September 2017

European Council (2007). EU Directive of the European Parliament and of the European Council on the estimation and management of flood risks (2007/60/EU).

Feder, J. (1988). Fractals. New York: Plenum Press.

Feng, Y., Liu, Y., Tong, X., Liu, M., & Deng, S. (2011). Modeling dynamic urban growth using cellular automata and particle swarm optimization rules. Landscape and Urban Planning, 102(3), 188-196.

Filho, B.S., Corradi, L.C., Cerqueira, G.C. &, Araújo W.L. (2003) Simulating the spatial patterns of change through the use of the DINAMICA model. XI Simpósio Brasileiro de Sensoriamento Remoto: 721–728

Filho BSS, H. O. Rodrigues a, W. L. S. Costa (2009) Modeling Environmental Dynamics with Dinamica EGO, Instituto de Geociencias - Centro de Sensoriamento Remoto, Av. Antonio Carlos, Universidade Federalde Minas Gerais - Campus Pampulha,

Forbes, D. (2011). Planning Sustainable Cities: Global Report on Human Settlements 2009 - By United Nations Human Settlements Programme. Geographical Research, 49(4), 447-448.

Forrester, J. W. (1969). Urban dynamics. Cambridge, MA: M.I.T. Press.

Fry, J. A., Coan, M. J., Homer, C. G., Meyer, D. K., & Wickham, J. D. (2009). Completion of the National Land Cover Database (NLCD) 1992-2001 land cover change retrofit product (No. 2008-1379). US Geological Survey.

Fuchs, R. J. (1994). Mega-city growth and the future. Tokyo: United Nations University Press.

Füssel, H. (2007, 08). Adaptation planning for climate change: Concepts, assessment approaches, and key lessons. Sustainability Science, 2(2), 265-275.

Gain, A., and Hoque, M. (2012, 11). Flood risk assessment and its application in the eastern part of Dhaka City, Bangladesh. Journal of Flood Risk Management, 6(3), 219-228.

Gandy, M. (2006). Planning, Anti-Planning, and the Infrastructure Crisis Facing Metropolitan Lagos. Cities in Contemporary Africa, 247-264.

Gardner, M. (1970). Mathematical games: The fantastic combinations of John Conway's new solitaire game "life". Scientific American, 223(4), 120-123.

Gaubatz, P. (1999). China's urban transformation: patterns and processes of morphological change in Beijing, Shanghai and Guangzhou. Urban studies, 36(9), 1495-1521.

Gersonius, B. (2012). The resilience approach to climate adaptation applied for flood risk. PhD Dissertation. Leiden: CRC/Balkema.

Ghosh, I., Hellweger, F. & Fritch, T. G. (2006) Fractal Generation of Artificial Sewer Networks for Hydrologic Simulations, in: Proceedings of the 2006 Esri International User Conference, San Diego, California, PP 1-25

Gong, Z., Tang, W., & Thill, J. C. (2012). Parallelization of ensemble neural networks for spatial land-use modeling. In Proceedings of the 5th ACM SIGSPATIAL International Workshop on Location-Based Social Networks, 48-54. ACM.

Goodacre, A., Bonham-Carter, G., Agterberg, P., & Wright, D. (1993, 10). A statistical analysis of the spatial association of seismicity with drainage patterns and magnetic anomalies in western Quebec. Tectonophysics, 225(4), 551.

Grinblat, Y., Gilichinsky, M. & Benenson, I. (2016). Cellular Automata Modeling of Land-Use/Land-Cover Dynamics: Questioning the Reliability of Data Sources and Classification Methods. Annals of the American Association of Geographers, 106(6), 1299-1320.

Haasnoot, M., Kwakkel, J. H., Walker, W. E., & Maat, J. T. (2013, 04). Dynamic adaptive policy pathways: A method for crafting robust decisions for a deeply uncertain world. Global Environmental Change, 23(2), 485-498.

Hammond M.J., Chen A.S., Djordjevic´ S., Butler D., Khan D.M., Rahman S.M.M. & Haque A.K.E. The development of a flood damage assessment tool for urban areas. 2012. In: Int. Conf. on Urban Drainage Modelling UDM 2012, Belgrade, Serbia.

Hagen, A. (2003) Fuzzy set approach to assessing similarity of categorical maps, International Journal of Geographic Information Science, 17:3, pp 235-249

Hall, J. W., Sayers, P. B., & Dawson, R. J. (2005). National-scale Assessment of Current and Future Flood Risk in England and Wales. Natural Hazards, 36(1-2), 147-164.

Hallegatte, S., Green, C., Nicholls, R. J., & Corfee-Morlot, J. (2013). Future flood losses in major coastal cities. Nature Climate Change, 3(9), 802-806.

Hallegatte, S., Ranger, N., Mestre, O., Dumas, P., Corfee-Morlot, J., Herweijer, C., & Wood, R. M. (2010, 12). Assessing climate change impacts, sea level rise and storm surge risk in port cities: A case study on Copenhagen. Climatic Change, 104(1), 113-137.

Han, H., Lai, S., Dang, A., Tan, Z., and Wu, C., (2009) Effectiveness of urban construction boundaries in Beijing: An assessment. Journal of Zhejiang University SCIENCE A, 10(9): 1285–1295

Haque A.K.E., Chakroborty S. & Zaman A.M. Development of Damage Functions for Dhaka City, Collaborative Research on Flood Resilience in Urban Areas (CORFU). 2014. Deliverable 3.4, Sec 4.4.2, (47-51).

Hasnat M.A. (2006) Flood disaster management in Dhaka city. In: K. Siddiqui & A.A. Hossain, eds. Options for flood risk and damage reduction in Bangladesh. Dhaka: University Press Limited, 121–129.

He, C., Okada, N., Zhang, Q., Shi, P. and Zhang, J (2006), Modeling urban expansion scenarios by coupling cellular automata model and system dynamic model in Beijing, China. Applied Geography, 26:3-4(323-345)

He, C., Okada, N., Zhang, Q., Shi, P. and Li, J. (2008) Modelling dynamic urban expansion processes incorporating a potential model with cellular automata. Landscape and Urban Planning, 86:1(79-91)

Hénonin, J., Hongtao, M., Zheng-Yu, Y., Hartnack, J., Havnø, K., Gourbesville, P., & Mark, O. (2013, 11). Citywide multi-grid urban flood modelling: The July 2012 flood in Beijing. Urban Water Journal, 12(1), 52-66.

Henze, M. (2008). Biological wastewater treatment: Principles, modelling and design. London: IWA Pub.

Herold, M., Couclelis, H. and Clarke, K. C. (2005) The role of spatial metrics in the analysis and modeling of urban land use change. Computers, Envirnoment and Urban Systems, 29:4 (369-399)

Herold, C., and Mouton, F. (2011). Global flood hazard mapping using statistical peak flow estimates. Hydrology and Earth System Sciences Discussions, 8(1), 305-363.

Hirabayashi, Y., Kanae, S., Emori, S., Oki, T., & Kimoto, M. (2008). Global projections of changing risks of floods and droughts in a changing climate. Hydrological Sciences Journal, 53(4), 754-772.

Homer C., Huang C., Yang L., Wylie B. & Coan M. Development of a 2001 National Landcover Database for the United States. Photogramm Eng Remote Sensing 2004, 70, (7), 829–840.

Homer, C. C. Huang, L. Yang, B. Wylie and Coan, M. (2004). Development of a 2001 National Landcover Database for the United States. Photogrammetric Engineering and Remote Sensing, Vol. 70, No. 7, 829-840

Horritt, M. S. and Bates, P. D. (2001). Predicting floodplain inundation: Raster-based modelling versus the finite-element approach. Hydrological Processes, 15(5), 825-842.

Horritt, M., & Bates, P. (2002, 11). Evaluation of 1D and 2D numerical models for predicting river flood inundation. Journal of Hydrology, 268(1-4), 87-99.

Hossain A.N.H. (2006) The impact of floods on Bangladesh and options for mitigation, an overview. In: K. Siddiqui & A.A. Hossain, eds. Options for flood risk and damage reduction in Bangladesh. Dhaka: University Press Limited, 55–70.

Hu, Z., & Lo, C. (2007). Modeling urban growth in Atlanta using logistic regression. Computers, environment and urban systems, 31(6), 667-688.

Huong, H. T., & Pathirana, A. (2013). Urbanization and climate change impacts on future urban flooding in Can Tho city, Vietnam. Hydrology and Earth System Sciences, 17(1), 379-394.

Huq, S., Kovats, S., Reid, H., & Satterthwaite, D. (2007). Editorial: Reducing risks to cities from disasters and climate change. Environment and Urbanization, 19(1), 3-15.

Intermap (2017) NEXTMap World10™, Filtered 10m resolution dtm with global coverage. URL[http://www.intermap.com/data/nextmap-world-10] Accessed September 2017

Islam A.K.M.S., Haque A. & Bala S.K. (2008) Hydrological Aspects of Flood 2007. Final Report, Institute of Water and Flood Management.

Islam K.M.N. (2005) Flood loss potentials in non-agricultural sectors: assessment methods and standard loss database for Bangladesh. Dhaka, Bangladesh: Palok Publishers.

Islam K.N. (2006) Impacts of flood in urban Bangladesh: micro and macro level analysis. Dhaka: AH Development Publishing House, 2006.

Iwugo KO, D'Arcy B. & Andoh R (2003) Aspects of land-based pollution of an African coastal megacity of Lagos, 122-124.

Jha, A. K., Bloch, R., & Lamond, J. (2012). Cities and flooding: A guide to integrated urban flood risk management for the 21st century. Washington, D.C.: World Bank.

Jongman, B., Ward, P. J., & Aerts, J. C. (2012). Global exposure to river and coastal flooding: Long term trends and changes. Global Environmental Change, 22(4), 823-835.

Kennedy, C. and Hoornweg, D. (2012) Mainstreaming Urban Metabolism. Journal of Industrial Ecology 16 (6), 780-782

Khan, D. M., Veerbeek, W., Chen, A. S., Hammond, M. J., Islam, F., Pervin, I., . . . Butler, D. (2015, 10). Back to the future: Assessing the damage of 2004 DHAKA FLOOD in The 2050 urban environment. Journal of Flood Risk Management J. Flood Risk Managent.

Kombe, W. J. (2005). Land use dynamics in peri-urban areas and their implications on the urban growth and form: The case of Dar es Salaam, Tanzania. Habitat International, 29(1), 113-135.

Kuang, W. (2012). Evaluating impervious surface growth and its impacts on water environment in Beijing-Tianjin-Tangshan Metropolitan Area. Journal of Geographical Sciences, 22(3), 535-547.

Kundzewicz, Z. W., Kanae, S., Seneviratne, S. I., Handmer, J., Nicholls, N., Peduzzi, P., . . . Sherstyukov, B. (2013, 12). Flood risk and climate change: Global and regional perspectives. Hydrological Sciences Journal, 59(1), 1-28.

Kurzbach, S., Hammond, M., Mark, O., Djordjevic, S., Butler, D., Gourbesville, P., ... & Chen, A. S. (2013). The development of socio-economic scenarios for urban flood risk management. Proceedings of the 8th International Novatech, Lyon, 23-27.

Landis, J., and Zhang, M. (1998). The second generation of the California urban futures model. Part 1: Model logic and theory. Environment and Planning B: Planning and Design, 25(5), 657-666.

Lau C.L., Smythe L.D., Craig S.B. & Weinstein P. (2010) Climate change, flooding, urbanisation and leptospirosis: fuelling the fire? Trans R Soc Trop Med Hyg 104, 631–638

Lee, D. B. (1994). Retrospective on Large-Scale Urban Models. Journal of the American Planning Association, 60(1), 35-40.

Leone, F., Lavigne, F., Paris, R., Denain, J., & Vinet, F. (2011, 01). A spatial analysis of the December 26th, 2004 tsunami-induced damages: Lessons learned for a better risk assessment integrating buildings vulnerability. Applied Geography, 31(1), 363-375.

Levy, J. M. (2015). Contemporary urban planning. Routledge.

Li, W., Ouyang, Z., Zhou, W., & Chen, Q. (2011, 08). Effects of spatial resolution of remotely sensed data on estimating urban impervious surfaces. Journal of Environmental Sciences, 23(8), 1375-1383.

Li, X., and Yeh, A. G. (2000, 03). Modelling sustainable urban development by the integration of constrained cellular automata and GIS. International Journal of Geographical Information Science, 14(2), 131-152.

Li, X., and Yeh, A. G. (2002, 06). Neural-network-based cellular automata for simulating multiple land use changes using GIS. International Journal of Geographical Information Science, 16(4), 323-343.

Li, X., Lin J. Y., Chen Y. M., Liu X. P. & Ai, B. (2013) Calibrating cellular automata based on landscape metrics by using genetic algorithms, International Journal of Geographical Information Science, 27, 594-613

Li, X., Yang, Q. S., & Liu, X. P. (2007). Genetic algorithms for determining the parameters of cellular automata in urban simulation. SCIENCE IN CHINA SERIES D EARTH SCIENCES-ENGLISH EDITION-, 50(12), 1857.

Li, X. and Yeh, A. (2004). Data mining of cellular automata's transition rules.International Journal of Geographical Information Science, 18, pp. 723–744.

Liu, Y. (2009) Modeling urban development with geographical information systems and cellular automata, CRC Press, Boca Raton, USA

Liu, Y. and Phinn, S. R. (2003). Modelling urban development with cellular automata incorporating fuzzy-set approaches. Computers, Environment, and Urban Systems 27: 637–58.

Long, Y. Mao, Q. and Dang, A (2009) Beijing Urban Development Model: Urban Growth Analysis and Simulation. Tsinghua Science and Technology, 14:6(782-794)

Long, Y., Shen, Z., Mao, Q. & Du, L. (2012). A Challenge to Configure Form Scenarios for Urban Growth Simulations Reflecting the Institutional Implications of Land-Use Policy in Geospatial Techniques in Urban Planning, Advances in Geographic Information Science, Springer Berlin Heidelberg

Lowry, I. S. (1964) A model of metropolis. Santa Monica, CA. The RAND corporation: 136

Luers, A. L. (2005). The surface of vulnerability: An analytical framework for examining environmental change. Global Environmental Change, 15(3), 214-223.

Lugeri, N., Genovese, E., Lavalle, C., & De Roo, A. (2006). Flood risk in Europe: analysis of exposure in 13 Countries. European Commission Directorate-General Joint Research Centre, EUR22525 EN.

Luo, H., Luo, L., Huang, G., Liu, P., Li, J., Hu, S., . . . Huang, X. (2009). Total pollution effect of urban surface runoff. Journal of Environmental Sciences, 21(9), 1186-1193.

Mark, O., Apirumanekul, C., Kamal, M. M., & Praydal, G. (2001). Modelling of Urban Flooding in Dhaka City. Urban Drainage Modeling.

Mark, O., Jørgensen, C., Hammond, M., Khan, D., Tjener, R., Erichsen, A., & Helwigh, B. (2015). A new methodology for modelling of health risk from urban flooding exemplified by cholera - case Dhaka, Bangladesh. Journal of Flood Risk Management.

Mas, J. F., Kolb, M., Paegelow, M. Olmedo, M. T. C. and Houet, T. (2014) Inductive pattern-based land use/cover change models: A comparison of four software packages. Environmental Modelling & Software 51, 94-111

McCann, P. (2012). Urban and regional economics. Milton Park, Abingdon, Oxon: Routledge.

McGarigal, K., Cushman, S. A., Neel, M. C., & Ene, E. (2002). FRAGSTATS: spatial pattern analysis program for categorical maps.

Mcgarigal, K., & Marks, B. J. (1995). FRAGSTATS: Spatial pattern analysis program for quantifying landscape structure.

Meadows, D. H., Meadows, D. H., Randers, J., & Behrens III, W. W. (1972). The limits to growth: a report to the club of Rome (1972). Universe Books, New York.

Meadows, D., Randers, J., & Meadows, D. (2004). Limits to growth: The 30-year update. Chelsea Green Publishing.

Mens, M. J., Klijn, F., Bruijn, K. M., & Beek, E. V. (2011). The meaning of system robustness for flood risk management. Environmental Science & Policy, 14(8), 1121-1131.

Merz, B., Hall, J., Disse, M., & Schumann, A. (2010,). Fluvial flood risk management in a changing world. Natural Hazards and Earth System Science, 10(3), 509-527.

Merz B., Kreibich H., Schwarze R. & Thieken A. (2010 )Review article' Assessment of economic flood damage. Nat Hazards Earth Syst Sci 10, (8), 1697–1724.

Messner F., Penning-Rowsell E., Green C., Meyer V., Tunstall S. & van der Veen A. (2007) Evaluating flood damages: guidance and recommendations on principles and methods. Technical report T09-06-01,

Mikhailov V.N. & Dotsenko M.A. Peculiarities of the hydrological regime of the Ganges and Brahmaputra river mouth area. Water Resour 2006, 33, 353–373.

Millennium Development Goals Report 2010. (2010). Millennium Development Goals Report.

Milly, P. C., Wetherald, R. T., Dunne, K. A., & Delworth, T. L. (2002). Increasing risk of great floods in a changing climate. Nature, 415(6871), 514-517.

Minghong, C., Hongwei, F., Yi, Z., & Guojian, H. (2013, 01). Integrated Flood Management for Beiyun River, China. Journal of Hydrology and Hydromechanics, 61(3).

Ministry for Economic Affairs (2015) Master plan national capital integrated coastal development    http://en.ncicd.com/wp-content/uploads/2015/02/MP-final-NCICD-dec-low-res.pdf

Mitchell, J. K. (2003). European River Floods in a Changing World. Risk Analysis, 23(3), 567-574.

Mitchell, T. M. (1997). Machine Learning. New York: McGraw-Hill.

Moel, H. D., & Aerts, J. C. (2010, 12). Effect of uncertainty in land use, damage models and inundation depth on flood damage estimates. Natural Hazards, 58(1), 407-425.

Moel, H. D., Alphen, J. V., & Aerts, J. C. (2009). Flood maps in Europe – methods, availability and use. Natural Hazards and Earth System Science, 9(2), 289-301.

Moghadam, H. S., & Helbich, M. (2013). Spatiotemporal urbanization processes in the megacity of Mumbai, India: A Markov chains-cellular automata urban growth model. Applied Geography, 40, 140-149.

Montz B.E. & Tobin G.A. Livin' large with levees: lessons learned and lost. Nat Hazards Rev 2008, 9, (3), 150–157.

Mousa, A. A., El-Shorbangy, M. A. and El-Wahed, A. (2012) Local search based hybrid particle swarm optimization algorithm for multiobjective optimization. Swarm and Evolutionary Computation, Volume 3. pp 1-14

Munich Re (2005). Weather catastrophes and climate change. Münchener Rückversicherungs-Gesellschaft, München. Knowledge series ppp2-264

Nakićenović, N. (2000). Special report on emissions scenarios: A special report of Working Group III of the Intergovernmental Panel on Climate Change. Cambridge: Cambridge University Press.

National Statistics Bureau (2011) Communiqué of the National Bureau of Statistics of People's Republic of China on Major Figures of the 2010 Population Census (No. 2)". URL [http://www.stats.gov.cn/english/NewsEvents/201104/t20110428_26449.html] Accessed August 2017

Nekkaa, M. and Boughaci, D. (2012) Improving support vector machine using a stochasticl local search for classification in data mining. In (Huang, T., Zueng, Z., Li, C. and Leung, C. S. (eds) Procceedings part II of the 19th International Conference ICONIP, Doha, Qatar, November 12-15, pp 168-175.Nicholls, R.J., Hanson, S., Herweijer, C., Patmore, N., Hallegatte, S., Corfee-Morlot, J., Château, J., Muir-Wood, R. (2008). Ranking port cities with high exposure and vulnerability to climate extremes exposure estimates. Paris, France: OECD Pub.

Næsset, E. (1995, 04). Testing for marginal homogeneity of remote sensing classification error matrices with ordered categories. ISPRS Journal of Photogrammetry and Remote Sensing, 50(2), 30-36.

Nishat A., Reazuddin M., Amin R. & Khan A.R. (2000) The 1998 flood: impact on environment of Dhaka city. Dhaka, Bangladesh: Dept. Of Environment in Co-operation with IUCN Bangladesh

NLCD 2001 (2001) URL[http://www.epa.gov/mrlc/definitions.htm] Accessed April 2016

O'neill, R. V., Riitters, K., Wickham, J., & Jones, K. B. (1999). Landscape Pattern Metrics and Regional Assessment. Ecosystem Health, 5(4), 225-233.

O'Neill, B. C., Kriegler, E., Ebi, K. L., Kemp-Benedict, E., Riahi, K., Rothman, D. S., ... & Levy, M. (2017). The roads ahead: narratives for shared socioeconomic pathways describing world futures in the 21st century. Global Environmental Change, 42, 169-180.

OECD Urban Policy Reviews: China 2015. (2015). OECD Publications Centre.

Olsen, A. S., Zhou, Q., Linde, J. J., & Arnbjerg-Nielsen, K. (2015). Comparing methods of calculating expected annual damage in urban pluvial flood risk assessments. Water, 7(1), 255-270.

Ong Y. S. and Keane A. J. (2004). "Meta-Lamarckian learning in memetic algorithms". IEEE Transactions on Evolutionary Computation 8 (2): 99–110.

Otukei, J. R., & Blaschke, T. (2010). Land cover change assessment using decision trees, support vector machines and maximum likelihood classification algorithms. International Journal of Applied Earth Observation and Geoinformation, 12, S27-S31.

Pan, A., Hou, A., Tian, F., Ni, G., & Hu, H. (2012). Hydrologically Enhanced Distributed Urban Drainage Model and Its Application in Beijing City. Journal of Hydrologic Engineering, 17(6), 667-678.

Pappenberger, F., Dutra, E., Wetterhall, F., & Cloke, H. (2012). Deriving global flood hazard maps of fluvial floods through a physical model cascade. Hydrology and Earth System Sciences Discussions, 9(5), 6615-6647.

Pathirana, A., Denekew, H. B., Veerbeek, W., Zevenbergen, C., & Banda, A. T. (2014,

03). Impact of urban growth-driven landuse change on microclimate and extreme precipitation — A sensitivity study. Atmospheric Research, 138, 59-72.

Penning-Rowsell E., Johnson C., Tunstall S., Tapsell S., Morris J., Chatterton J. & Green C. (2005) The benefits of flood and coastal risk management: a manual of assessment techniques (the multicoloured manual). London: Flood Hazard Research Centre, Middlesex University

Phi, H. L. (2007). Climate change and urban flooding in Ho Chi Minh City. In Proceedings of Third International Conference on Climate and Water (pp. 194-199).

Potere, D., & Schneider, A. (2007). A critical look at representations of urban areas in global maps. GeoJournal, 69(1-2), 55-80.

Rabbani, A., Aghababaee, H., & Rajabi, M. A. (2012). Modeling dynamic urban growth using hybrid cellular automata and particle swarm optimization. Journal of Applied Remote Sensing, 6(1), 063582-1.

Rahman R., Haque A., Khan S.A., Salehin M. & Bala S.K. Investigation of hydrological aspects of flood-2004 with special emphasis on Dhaka city, Bangladesh. Dhaka: Institute of Water and Flood Management (IWFM), Bangladesh University of Engineering and Technology (BUET), 2005.

RAJUK (1997) Dhaka Metropolitan Development Plan (1995-2015): Volume I: Dhaka Structure Plan (1995-2015). Dhaka: RAJUK.

Reuters (2011) [URL: http://www.reuters.com/article/uk-china-floods-idUSL-NE75M05520110623] Accessed September 2017.

Roo, A. P., Wesseling, C. G., & Deursen, W. P. (2000). Physically based river basin modelling within a GIS: The LISFLOOD model. Hydrological Processes, 14(1112), 1981-1992.

Roo, A. D., Barredo, J., Lavalle, C., Bodis, K., & Bonk, R. (2007). Potential Flood Hazard and Risk Mapping at Pan-European Scale. Lecture Notes in Geoinformation and Cartography Digital Terrain Modelling, 183-202.

Roy, A. (2012). Urban Informality. Oxford Handbooks Online.

Rüther, H., Martine, H. M., & Mtalo, E. (2002). Application of snakes and dynamic programming optimisation technique in modeling of buildings in informal settlement areas. ISPRS Journal of Photogrammetry and Remote Sensing, 56(4), 269-282.

Schlitte F. (2013) Scenarios for long term economic city growth – models and results for selected CORFU case studies. In: D. Butler, A.S. Chen, S. Djordjevic & M.J. Hammond, eds. Urban flood resilience. Exeter, United Kingdom: University of Exeter, 2013, pp. 27–28. URL[http://corfu7.eu/media/universityofexeter/research/microsites/corfu/1publicdocs/ conferencepapers/ A2_246_Schlitte.

pdf] Accessed September 2016.

Scoones, I. (2004). Climate Change and the Challenge of Non-equilibrium Thinking. IDS Bulletin, 35(3), 114-119.

Schreider, S. Y., Smith, D. I., & Jakeman, A. J. (2000). Climate change impacts on urban flooding. Climatic Change, 47(1-2), 91-115.

Scussolini, P., Aerts, J. C., Jongman, B., Bouwer, L. M., Winsemius, H. C., Moel, H. D., & Ward, P. J. (2015). FLOPROS: An evolving global database of flood protection standards. Natural Hazards and Earth System Sciences Discussions, 3(12), 7275-7309.

Semadeni-Davies, A., Hernebring, C., Svensson, G., & Gustafsson, L. (2008,). The impacts of climate change and urbanisation on drainage in Helsingborg, Sweden: Combined sewer system. Journal of Hydrology, 350(1-2), 100-113.

Shan, J., Alkheder, S., & Wang, J. (2008). Genetic algorithms for the calibration of cellular automata urban growth modeling. Photogrammetric Engineering & Remote Sensing, 74(10), 1267-1277.

Shepherd, J. (2006). Evidence of urban-induced precipitation variability in arid climate regimes. Journal of Arid Environments, 67(4), 607-628.

Shi, P., Yuan, Y., Zheng, J., Wang, J., Ge, Y., & Qiu, G. (2007). The effect of land use/ cover change on surface runoff in Shenzhen region, China. Catena, 69(1), 31-35.

Shi, P., Yuan, Y., Zheng, J., Wang, J., Ge, Y., & Qiu, G. (2007). The effect of land use/ cover change on surface runoff in Shenzhen region, China. Catena, 69(1), 31-35.

Shuttle Radar Topography Mission (SRTM) (2016) 1 Arc-Second Global. (n.d.). R URL[https://lta.cr.usgs.gov/SRTM1Arc] Accessed April 2016

Skamarock, W.C., Klemp, J.B., Dudhia, J., Gill, D.O., Barker, D.M., Wang, W. & Powers, J.G. (2005) A Description of the Advanced Research WRF Version 2. Technical Report; National Center for Atmospheric Research; Boulder, Colorado, USA

Siddiqui K. and Hossain A.A. (2006) Options for flood risk and damage reduction in Bangladesh. Dhaka: University Press Limited.

Singh, V. P. (1992). Elementary hydrology. Englewood Cliffs, NJ: Prentice Hall.

Soares-Filho B, Rodrigues H, Costa W (2009) Modeling Environmental Dynamics With Dinamica EGO. Belo Horizonte, Brazil, Sensoriamento Remoto/Universidade Federal de Minas Gerais

Soil Conservation Service (SCS) (1956, 1964, 1971, 1972, 1985 ), 'Hydrology, National Engineering Handbook, Supplement A, Section 4, Chapter 10, Soil Conservation Service, U.S.D.A., Washington, D.C.

Solomon, S., Qin, D., Manning, M., Chen, Z., Marquis, M., Averyt, K.B., Tignor, M., Miller, H.L., eds. (2005). Climate Change 2005-The Physical Science Basis: Working Group I Contribution to the Fourth Assessment Report of the IPCC. Cambridge, UK and New York, NY, USA: Cambridge University Press.

State of the world's cities, 2008/2009: Harmonious cities. (2009). Choice Reviews Online, 46(08).

State of world population 2007: Unleashing the Potential of Urban Growth. (2007). New York: UNFPA.

Stern, N. H. (2007). The economics of climate change: The Stern review. Cambridge, UK: Cambridge University Press.

Straatman, B., White, R., & Engelen, G. (2004). Towards an automatic calibration procedure for constrained cellular automata. Computers, Environment and Urban Systems, 28(1-2), 149-170.

Suppasri, A., Shuto, N., Imamura, F., Koshimura, S., Mas, E., & Yalciner, A. C. (2012). Lessons Learned from the 2011 Great East Japan Tsunami: Performance of Tsunami Countermeasures, Coastal Buildings, and Tsunami Evacuation in Japan. Pure and Applied Geophysics Pure Appl. Geophys., 170(6-8), 993-1018.

Tan, Q., Liu, Z., & Li, X. (2008). Analysis of the Degree of Urban Impervious Surface based on Object-Oriented Method. IGARSS 2008 - 2008 IEEE International Geoscience and Remote Sensing Symposium.

Tawhid, K. G. (2004). Causes and effects of water logging in Dhaka City, Bangladesh. TRITA-LWR master thesis, Department of Land and Water Resource Engineering, Royal Institute of Technology, Stockholm.

Thapa, R. B., & Murayama, Y. (2011). Urban growth modeling of Kathmandu metropolitan region, Nepal. Computers, Environment and Urban Systems, 35(1), 25-34.

The Economist (2016) A Summary of the Liveability Ranking and Overview. A report by the Economist Inteligence Unit. Retrieved from: http://pages.eiu.com/rs/783-XMC-194/images/Liveability_ August2016.pdf

Tobin G.A. (1995) The Levee love affair: a stormy relationship. Water Resour Bull 31, (3), 359–367.

Trends in urbanization. (2014). Statistical Papers - United Nations (Ser. A), Population and Vital Statistics Report World Urbanization Prospects, 7-12.

Turing, A. M. (1936). On computable numbers, with an application to the Entscheidungsproblem. J. of Math, 58(345-363), 5.

Turner, M. G. (1990). Spatial and temporal analysis of landscape patterns. Landscape Ecology, 4(1), 21-30.

Turner, M. G. (1988). A spatial simulation model of land use changes in a Piedmont county in Georgia. Applied Mathematics and Computation, 27(1), 39-51.

UN (2007) State of world population 2007: Unleashing the Potential of Urban Growth. New York: UNFPA.

UN (2012) World urbanization prospects the 2011 revision. New York: United Nations, Department of Economic and Social Affairs, Population Division

UN (2015). United Nations, Department of Economic and Social Affairs, Population Division World Urbanization Prospects: The 2014 Revision, CD-ROM Edition. 2014.

UN Habitat. Sustainable urbanization: Local actions for urban poverty reduction, emphasis on finance and planning. 2007. 21st Session of the Governance Council, 16–20 April, Nairobi, Kenya, 2007.

United Nations Human Settlements Programme, Planning Sustainable Cities: Global Report on Human Settlements 2009 (UN HABITAT and Earthscan), 2009. (2011). Environment and Urbanization Asia, 2(1), 149-149.

USGS and NASA (2009) Global Land Survey 2005, Sioux Falls, SD USA: USGS Center for Earth Resources Observation and Science (EROS)

Veerbeek, W., & Zevenbergen, C. (2009). Deconstructing urban flood damages: Increasing the expressiveness of flood damage models combining a high level of detail with a broad attribute set. Journal of Flood Risk Management, 2(1), 45-57.

Veerbeek, W., Schitte, F. and Zevenbergen, C. (2014) Economic an urban growth models. Colaborative research on flood resilience in urban areas. Technical report D1.3 URL[http://corfu-fp7.eu/] accessed April 2014

Veerbeek, W., Ashley, R. M., Zevenbergen, C., Rijke, J. S., & Gersonius, B. (2010, December). Building adaptive capacity for flood proofing in urban areas through synergistic interventions. In ICSU 2010: Proceedings of the First International Conference on Sustainable Urbanization, 15-17 December 2010, Hong Kong, China. Hong Kong Polytechnic University.

Veerbeek, W., Pathirana, A., Ashley, R., & Zevenbergen, C. (2015). Enhancing the calibration of an urban growth model using a memetic algorithm. Computers, Environment and Urban Systems, 50, 53-65.

Veerbeek, W., and Zevenbergen, C. (2009). Deconstructing urban flood damages: Increasing the expressiveness of flood damage models combining a high level of detail with a broad attribute set. Journal of Flood Risk Management, 2(1), 45-57.

Veerbeek, W., Gersonius, B., Chen, A. S., Hammond, M. J., Khan, D. M., Radhakrishnan, M., & Zevenbergen, C. (2014) Applied Flood Resiliency: A Method for Determining the Recovery Capacity for Fast Growing Mega Cities.13th International Conference on Urban Drainage, Sarawak, Malaysia, 7-12 September

Veerbeek W., Denekew, H.B., Pathirana, A., Bacchin, T., Brdjanovic, D. (2011) Urban growth modeling to predict the changes in the urban microclimate and urban water cycle. In: 12th International Conference on Urban Drainage Modelling. Brazil.

Veerbeek W., Pathirana A., Mudenda S.H. & Brdjanovic D. (2012) Application of urban growth model to project slum development and its implications on water supply and sanitation planning. In Proceedings 9th International Conference on Urban Drainage Modelling (pp. 3–7).

Vojinović, Z. (2015). Flood risk: The holistic perspective: From integrated to interactive planning for flood resilience. London, UK: IWA Publishing.

Volp, N. D., Prooijen, B. C., & Stelling, G. S. (2013). A finite volume approach for shallow water flow accounting for high-resolution bathymetry and roughness data. Water Resources Research, 49(7), 4126-4135.

Von Neumann, J. (1951). The general and logical theory of automata. Cerebral mechanisms in behavior, 1(41), 1-2

Walker, W., Haasnoot, M., & Kwakkel, J. (2013). Adapt or Perish: A Review of Planning Approaches for Adaptation under Deep Uncertainty. Sustainability, 5(3), 955-979.

Watkins, R., & Kolokotroni, M. (2012, November). The London Urban Heat Island–upwind vegetation effects on local temperatures. In PLEA2012-28th Conference, Lima, Perú November 2012.

Weber A., (1909). Über den Standort der Industrien, Tübingen, J.C.B. Mohr. English translation: The Theory of the Location of Industries, Chicago, Chicago University Press, 1929.

Weng, Q. (2001, 12). Modeling Urban Growth Effects on Surface Runoff with the Integration of Remote Sensing and GIS. Environmental Management, 28(6), 737-748.

White, R., & Engelen, G. (2000). High-resolution integrated modelling of the spatial dynamics of urban and regional systems. Computers, Environment and Urban Systems, 24(5), 383-400.

White, R. (1998). Cities and cellular automata. Discrete Dynamics in Nature and Society, 2(2), 111-125.

White, R. and Engelen, G. (1993) Cellular automata and fractal urban form: a cellular modelling approach to the evolution of urban land-use patterns. Environment and Planning A. 02/1993; 25(8):1175-1199.

Winsemius, H. C., Beek, L. P., Jongman, B., Ward, P. J., & Bouwman, A. (2013). A framework for global river flood risk assessments. Hydrol. Earth Syst. Sci. Hydrology and Earth System Sciences, 17(5), 1871-1892.

Wolfram, S. (2002). A new kind of science. Champaign, IL: Wolfram Media.

Woodhead, S., Asselman, N., Zech, Y., Soares-Fraza~o, S., Bates, P., Kortenhaus, A., (2007). Evaluation of Inundation Models. Limits and Capabilities of Models, FLOODsite, T08-07-01.

Yang, Q., Li, X., & Shi, X. (2008). Cellular automata for simulating land use changes based on support vector machines. Computers & Geosciences, 34(6), 592-602.

Yang, X., & Liu, Z. (2005). Use of satellite-derived landscape imperviousness index to characterize urban spatial growth. Computers, Environment and Urban Systems, 29(5), 524-540.

Yang, J. and Jinxing, Z. (2007) The failure and success of greenbelt program in Beijing, in: Urban Forestry & Urban Greening, 6:4(287-296)

Yi, W., Gao, Z. and Chen, M. (2012) Dynamic modelling of future land-use change: a comparison between CLUE-S and Dinamica EGO models, Proc. SPIE 8513, Remote Sensing and Modeling of Ecosystems for Sustainability IX, 85130H, October 24, San Diego, California, USA

Zevenbergen, C. (2011). Urban flood management. Boca Raton: CRC Press.

Zhai, P., Zhang, X., Wan, H., & Pan, X. (2005). Trends in Total Precipitation and Frequency of Daily Precipitation Extremes over China. Journal of Climate, 18(7), 1096-1108.

Zhao, P. (2010) Sustainable urban expansion and transportation in a growing megacity: Consequences of urban sprawl for mobility on the urban fringe of Beijing, in: Habitat International, 34:2(236-243)

Zhao, L. (2016). China's 13th Five-Year Plan: Road Map for Social Development. East Asian Policy, 08(03), 19-32.

Zhang, C. and Ma, Y. (2012) Ensemble Machine Learning, Methods and applications. Springer, New York, USA

Zhang, Z., Yang, H., & Shi, M. (2011). Analyses of water footprint of Beijing in an interregional input–output framework. Ecological Economics, 70(12), 2494-2502.

Zhou, Y., Zhao, J., Liu, Z., Li, W., Zhang, J., & Tang, Y. (2013) Flood control of an overpass in downtown Beijing. International Conference on Flood Resilience: Experiences in Asia and Europe. 5-7 September 2013, Exeter, United Kingdom

# Appendix A: Urban growth and riverine flooding

# A1 Land Use and Land Cover Classification

The land use and land cover classification used in the maps reflecting the urban growth scenarios is based on the NLCD 2001 Land Cover Class Definitions (Homer et al, 2004). The following classes were used:

| Code | LULC class |
| --- | --- |
| 11 | OPEN WATER |
| 22 | DEVELOPED, LOW INTENSITY |
| 23 | DEVELOPED, MEDIUM INTENSITY |
| 24 | DEVELOPED, HIGH INTENSITY |
| 31 | BARREN LAND |
| 41 | DECIDUOUS FOREST |
| 42 | EVERGREEN FOREST |
| 43 | MIXED FOREST |
| 52 | SHRUB |
| 71 | GRASSLAND/HERBACEOUS |
| 90 | WOODY WETLANDS |
| 95 | EMERGENT HERBACEOUS WETLANDS |

# A2 Urban Growth 2010-2060

The maps show the estimated urban footprint for the baseyear 2010 (dark grey) and the proejcted footprint for 2060 (light grety). All cities are represented at the same scale

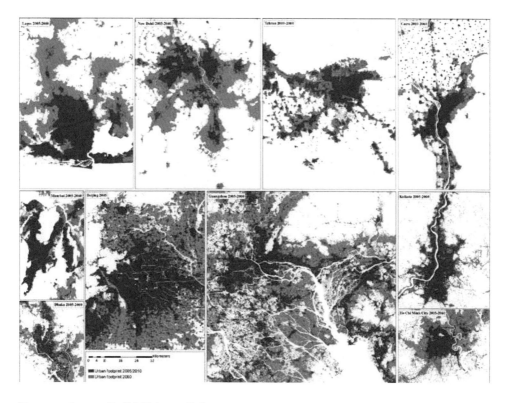

Top row: Lagos, Delhi, Tehran, Cairo
Bottom row: Mumbai/Dhaka, Beijing, Guangzhou-Shenzhen, Calcutta/Ho Chi Minh

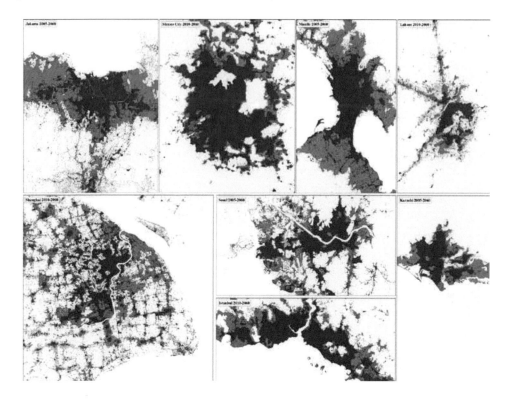

Top row: Jakarta, Mexico City, Manila, Lahore
Bottom row: Shanghai, Seoul/Istanbul, Karachi

# A3 Urban Flood Extent

The tables in this appendix provide an overview of the estimated average urban flood extent for the two base years (e.g. 1995 and 2005) as well as for the range of future projections until 2060. The estimations are shown for the initial setup using based on the GLOFRIS model and the developed LULC-maps, as well as for the two validation methods (Upscaled and Downscaled). The final outcomes used in the assessment are based on the Average values. Finally the tables show the resulting rank based on the size of the urban flood extent for 2015 and 2060.

| Beijing | 1995 | 2005 | 2015 | 2025 | 2035 | 2045 | 2055 | 2060 | Avg. Dev. | rank 2015 | rank 2060 |
|---|---|---|---|---|---|---|---|---|---|---|---|
| GLOFRIS | 92.6 | 112.8 | 138.4 | 178.3 | 226.3 | 278.7 | 327.9 | 355.3 | | 5 | 4 |
| Upscaled | 90.5 | 113.8 | 132.8 | 174.6 | 220.4 | 274.5 | 322.5 | 348.5 | 2.1% | 4 | 5 |
| Downscaled | 84.1 | 103.1 | 126.3 | 162.5 | 205.9 | 253.1 | 296.1 | 320.5 | 9.1% | 3 | 3 |
| Average | 89.1 | 109.9 | 132.5 | 171.8 | 217.5 | 268.8 | 315.5 | 341.4 | | 4 | 4 |

| Cairo | 1990 | 2010 | 2015 | 2025 | 2035 | 2045 | 2055 | 2060 | Avg. Dev. | rank 2015 | rank 2060 |
|---|---|---|---|---|---|---|---|---|---|---|---|
| GLOFRIS | 31.0 | 49.6 | 56.6 | 70.3 | 84.4 | 99.1 | 112.7 | 119.1 | | 10 | 12 |
| Upscaled | 25.7 | 41.9 | 47.8 | 59.7 | 71.9 | 82.7 | 92.3 | 98.2 | 16.3% | 12 | 12 |
| Downscaled | 22.1 | 37.8 | 43.8 | 56.1 | 69.3 | 83.1 | 96.0 | 102.0 | 19.8% | 11 | 11 |
| **Average** | **26.2** | **43.1** | **49.4** | **62.0** | **75.2** | **88.3** | **100.3** | **106.4** | | **11** | **12** |

| Calcutta | 1995 | 2010 | 2015 | 2025 | 2035 | 2045 | 2055 | 2060 | Avg. Dev. | rank 2015 | rank 2060 |
|---|---|---|---|---|---|---|---|---|---|---|---|
| GLOFRIS | 72.5 | 181.2 | 211.9 | 241.6 | 300.8 | 362.0 | 421.9 | 452.3 | | 3 | 3 |
| Upscaled | 68.0 | 155.7 | 184.3 | 204.5 | 248.4 | 294.8 | 338.7 | 361.7 | 15.5% | 3 | 3 |
| Downscaled | 68.5 | 170.4 | 198.4 | 225.3 | 278.7 | 332.8 | 385.6 | 411.9 | 7.2% | 2 | 2 |
| **Average** | **69.7** | **169.1** | **198.2** | **223.8** | **276.0** | **329.9** | **382.1** | **408.6** | | **3** | **2** |

| Delhi | 2000 | 2010 | 2015 | 2025 | 2035 | 2045 | 2055 | 2060 | Avg. Dev. | rank 2015 | rank 2060 |
|---|---|---|---|---|---|---|---|---|---|---|---|
| GLOFRIS | 69.0 | 111.1 | 135.7 | 183.7 | 231.8 | 275.5 | 312.8 | 331.3 | | 6 | 5 |
| Upscaled | 54.6 | 90.3 | 111.2 | 142.6 | 178.4 | 214.5 | 240.6 | 255.5 | 21.4% | 6 | 6 |
| Downscaled | 49.6 | 83.7 | 104.5 | 146.2 | 188.9 | 228.3 | 262.2 | 279.1 | 20.5% | 5 | 4 |
| **Average** | **57.7** | **95.0** | **117.1** | **157.5** | **199.7** | **239.4** | **271.9** | **288.6** | | **6** | **5** |

| Dhaka | 2000 | 2010 | 2015 | 2025 | 2035 | 2045 | 2055 | 2060 | Avg. Dev. | rank 2015 | rank 2060 |
|---|---|---|---|---|---|---|---|---|---|---|---|
| GLOFRIS | 25.8 | 47.3 | 60.1 | 106.0 | 143.1 | 179.5 | 213.0 | 228.8 | | 9 | 7 |
| Upscaled | 17.3 | 53.4 | 72.8 | 138.6 | 201.7 | 266.8 | 331.9 | 359.7 | 37.5% | 10 | 4 |
| Downscaled | 24.9 | 46.0 | 58.6 | 103.8 | 140.2 | 176.0 | 209.0 | 224.5 | 2.3% | 7 | 5 |
| **Average** | **22.7** | **48.9** | **63.8** | **116.1** | **161.6** | **207.4** | **251.3** | **271.0** | | **9** | **6** |

| Guangzhou | 1990 | 2005 | 2015 | 2025 | 2035 | 2045 | 2055 | 2060 | Avg. Dev. | rank 2015 | rank 2060 |
|---|---|---|---|---|---|---|---|---|---|---|---|
| GLOFRIS | 102.9 | 471.4 | 661.3 | 840.5 | 1007.4 | 1160.1 | 1284.8 | 1346.7 | | 1 | 1 |
| Upscaled | 118.7 | 523.9 | 792.4 | 1040.1 | 1273.9 | 1471.7 | 1681.2 | 1771.4 | 23.2% | 1 | 1 |
| Downscaled | 84.3 | 389.2 | 550.1 | 702.1 | 843.5 | 974.2 | 1080.8 | 1133.9 | 16.6% | 1 | 1 |
| Average | 102.0 | 461.5 | 668.0 | 860.9 | 1041.6 | 1202.0 | 1348.9 | 1417.4 | | 1 | 1 |

| Ho Chi Minh City | 1990 | 2005 | 2015 | 2025 | 2035 | 2045 | 2055 | 2060 | Avg. Dev. | rank 2015 | rank 2060 |
|---|---|---|---|---|---|---|---|---|---|---|---|
| GLOFRIS | 6.3 | 30.0 | 53.8 | 84.1 | 119.2 | 159.0 | 201.8 | 224.2 | | 12 | 8 |
| Upscaled | 8.3 | 27.5 | 62.4 | 86.4 | 116.7 | 153.4 | 196.4 | 215.9 | 8.8% | 11 | 8 |
| Downscaled | 4.0 | 11.0 | 16.4 | 25.8 | 42.5 | 66.1 | 93.1 | 107.5 | 58.4% | 14 | 10 |
| Average | 6.2 | 22.9 | 44.2 | 65.4 | 92.8 | 126.2 | 163.8 | 182.5 | | 12 | 8 |

| Istanbul | 1990 | 2010 | 2015 | 2025 | 2035 | 2045 | 2055 | 2060 | Avg. Dev. | rank 2015 | rank 2060 |
|---|---|---|---|---|---|---|---|---|---|---|---|
| GLOFRIS | 0.2 | 1.2 | 1.6 | 2.0 | 2.9 | 3.6 | 4.7 | 5.1 | | 18 | 18 |
| Upscaled | 0.7 | 1.4 | 1.4 | 1.4 | 4.2 | 4.2 | 6.3 | 8.7 | 59.3% | 18 | 17 |
| Downscaled | 0.0 | 0.0 | 0.0 | 0.0 | 0.0 | 0.0 | 0.0 | 0.0 | 99.9% | 18 | 18 |
| Average | 0.3 | 0.9 | 1.0 | 1.2 | 2.4 | 2.6 | 3.7 | 4.6 | | 18 | 18 |

| Jakarta | 1995 | 2005 | 2015 | 2025 | 2035 | 2045 | 2055 | 2060 | Avg. Dev. | rank 2015 | rank 2060 |
|---|---|---|---|---|---|---|---|---|---|---|---|
| GLOFRIS | 22.9 | 34.9 | 73.8 | 112.0 | 148.5 | 195.4 | 236.9 | 257.6 | | 8 | 6 |
| Upscaled | 20.8 | 33.6 | 82.7 | 111.8 | 139.4 | 187.1 | 230.7 | 253.2 | 5.0% | 7 | 7 |
| Downscaled | 13.9 | 23.0 | 53.4 | 78.9 | 97.2 | 115.7 | 131.0 | 137.6 | 37.1% | 8 | 8 |
| Average | 19.2 | 30.5 | 70.0 | 100.9 | 128.3 | 166.1 | 199.5 | 216.1 | | 8 | 7 |

| Karachi | 1990 | 2005 | 2015 | 2025 | 2035 | 2045 | 2055 | 2060 | Avg. Dev. | rank 2015 | rank 2060 |
|---|---|---|---|---|---|---|---|---|---|---|---|
| GLOFRIS | 2.3 | 6.2 | 8.2 | 9.5 | 11.7 | 17.7 | 18.9 | 19.7 | | 16 | 16 |
| Upscaled | 2.1 | 5.7 | 6.2 | 7.2 | 9.7 | 15.6 | 17.3 | 18.5 | 13.7% | 16 | 16 |
| Downscaled | 2.0 | 5.4 | 7.4 | 8.7 | 10.8 | 16.7 | 17.9 | 18.7 | 8.4% | 16 | 16 |
| Average | 2.1 | 5.7 | 7.3 | 8.4 | 10.8 | 16.7 | 18.0 | 19.0 | | 16 | 16 |

| Lagos | 2000 | 2010 | 2015 | 2025 | 2035 | 2045 | 2055 | 2060 | Avg. Dev. | rank 2015 | rank 2060 |
|---|---|---|---|---|---|---|---|---|---|---|---|
| GLOFRIS | 37.8 | 60.5 | 78.6 | 103.8 | 124.8 | 148.4 | 172.7 | 182.9 | | 7 | 10 |
| Upscaled | 34.4 | 62.5 | 78.3 | 102.1 | 122.1 | 146.4 | 175.3 | 183.2 | 2.4% | 8 | 9 |
| Downscaled | 33.4 | 56.8 | 69.5 | 92.9 | 112.3 | 134.2 | 157.1 | 166.7 | 9.6% | 6 | 7 |
| Average | 35.2 | 59.9 | 75.5 | 99.6 | 119.7 | 143.0 | 168.4 | 177.6 | | 7 | 10 |

| Lahore | 1990 | 2010 | 2015 | 2025 | 2035 | 2045 | 2055 | 2060 | Avg. Dev. | rank 2015 | rank 2060 |
|---|---|---|---|---|---|---|---|---|---|---|---|
| GLOFRIS | 2.7 | 13.7 | 16.3 | 23.8 | 32.6 | 42.6 | 54.4 | 60.8 | | 15 | 14 |
| Upscaled | 0.5 | 6.1 | 9.0 | 13.2 | 19.1 | 27.2 | 37.1 | 43.0 | 45.4% | 15 | 15 |
| Downscaled | 1.7 | 10.6 | 12.7 | 18.9 | 26.4 | 35.2 | 45.5 | 51.3 | 21.2% | 15 | 14 |
| Average | 1.7 | 10.1 | 12.7 | 18.6 | 26.0 | 35.0 | 45.7 | 51.7 | | 15 | 15 |

| Manila | 1990 | 2005 | 2015 | 2025 | 2035 | 2045 | 2055 | 2060 | Avg. Dev. | rank 2015 | rank 2060 |
|---|---|---|---|---|---|---|---|---|---|---|---|

| | | | | | | | | | Avg. Dev. | rank 2015 | rank 2060 |
|---|---|---|---|---|---|---|---|---|---|---|---|
| GLOFRIS | 19.6 | 49.2 | 55.6 | 68.9 | 84.1 | 101.0 | 119.8 | 128.4 | | 11 | 11 |
| Upscaled | 10.4 | 27.2 | 31.9 | 44.2 | 56.5 | 72.7 | 80.5 | 85.7 | 37.1% | 13 | 13 |
| Downscaled | 15.4 | 38.6 | 44.0 | 55.3 | 68.8 | 84.1 | 101.0 | 108.9 | 18.6% | 10 | 9 |
| **Average** | **15.2** | **38.4** | **43.9** | **56.1** | **69.8** | **85.9** | **100.4** | **107.7** | | **13** | **11** |

| **Mexico City** | 1990 | 2010 | 2015 | 2025 | 2035 | 2045 | 2055 | 2060 | Avg. Dev. | rank 2015 | rank 2060 |
|---|---|---|---|---|---|---|---|---|---|---|---|
| GLOFRIS | 90.7 | 130.0 | 138.6 | 152.2 | 164.8 | 178.2 | 187.9 | 193.3 | | 4 | 9 |
| Upscaled | 81.4 | 118.0 | 125.2 | 136.8 | 145.3 | 157.9 | 170.7 | 173.5 | 10.2% | 5 | 10 |
| Downscaled | 76.3 | 112.4 | 120.6 | 133.5 | 145.6 | 158.4 | 167.7 | 173.0 | 12.3% | 4 | 6 |
| **Average** | **82.8** | **120.1** | **128.1** | **140.8** | **151.9** | **164.8** | **175.4** | **179.9** | | **5** | **9** |

| **Mumbai** | 1990 | 2005 | 2015 | 2025 | 2035 | 2045 | 2055 | 2060 | Avg. Dev. | rank 2015 | rank 2060 |
|---|---|---|---|---|---|---|---|---|---|---|---|
| GLOFRIS | 11.6 | 19.8 | 24.9 | 32.0 | 40.2 | 48.4 | 55.8 | 59.0 | | 14 | 15 |
| Upscaled | 14.4 | 19.1 | 20.4 | 26.4 | 37.8 | 48.3 | 50.9 | 52.1 | 11.2% | 14 | 14 |
| Downscaled | 10.9 | 18.7 | 23.7 | 30.8 | 38.9 | 47.0 | 54.1 | 57.2 | 4.1% | 13 | 13 |
| **Average** | **12.3** | **19.2** | **23.0** | **29.7** | **39.0** | **47.9** | **53.6** | **56.1** | | **14** | **14** |

| **Seoul** | 1990 | 2005 | 2015 | 2025 | 2035 | 2045 | 2055 | 2060 | Avg. Dev. | rank 2015 | rank 2060 |
|---|---|---|---|---|---|---|---|---|---|---|---|
| GLOFRIS | 30.1 | 37.7 | 46.9 | 54.4 | 59.7 | 63.3 | 66.1 | 67.3 | | 13 | 13 |
| Upscaled | 45.6 | 67.5 | 72.9 | 81.6 | 90.2 | 99.1 | 105.5 | 109.0 | 58.2% | 9 | 11 |
| Downscaled | 28.9 | 36.3 | 45.4 | 52.7 | 57.9 | 61.4 | 64.2 | 65.3 | 3.2% | 9 | 12 |
| **Average** | **34.9** | **47.1** | **55.1** | **62.9** | **69.3** | **74.6** | **78.6** | **80.5** | | **10** | **13** |

| **Shanghai** | 1990 | 2010 | 2015 | 2025 | 2035 | 2045 | 2055 | 2060 | Avg. Dev. | rank 2015 | rank 2060 |
|---|---|---|---|---|---|---|---|---|---|---|---|
| GLOFRIS | 97.3 | 291.3 | 335.8 | 399.1 | 444.5 | 480.1 | 511.7 | 525.9 | | 2 | 2 |
| Upscaled | 73.0 | 243.6 | 295.0 | 345.6 | 401.0 | 434.8 | 462.0 | 476.0 | 13.2% | 2 | 2 |
| Downscaled | 15.1 | 22.8 | 26.9 | 31.0 | 34.2 | 36.8 | 38.9 | 39.8 | 91.3% | 12 | 15 |
| **Average** | **61.8** | **185.9** | **219.2** | **258.5** | **293.3** | **317.2** | **337.5** | **347.3** | | **2** | **3** |

| **Tehran** | 1990 | 2010 | 2015 | 2025 | 2035 | 2045 | 2055 | 2060 | Avg. Dev. | rank 2015 | rank 2060 |
|---|---|---|---|---|---|---|---|---|---|---|---|
| GLOFRIS | 0.8 | 5.5 | 6.4 | 8.4 | 10.4 | 12.2 | 13.4 | 14.3 | | 17 | 17 |
| Upscaled | 0.0 | 3.4 | 3.8 | 4.8 | 6.3 | 6.5 | 6.6 | 7.5 | 50.8% | 17 | 18 |
| Downscaled | 0.6 | 4.5 | 5.3 | 7.2 | 9.0 | 10.5 | 11.5 | 12.2 | 15.7% | 17 | 17 |
| **Average** | **0.5** | **4.5** | **5.2** | **6.8** | **8.6** | **9.7** | **10.5** | **11.3** | | **17** | **17** |

# A4 Risk profiles megacities

## Table of Content

| | |
|---|---:|
| Beijing | 258 |
| Cairo | 262 |
| Calcutta | 266 |
| Delhi | 270 |
| Dhaka | 274 |
| Guangzhou-Shenzhen | 278 |
| Ho Chi Minh City | 282 |
| Istanbul | 286 |
| Jakarta | 290 |
| Karachi | 294 |
| Lagos | 298 |
| Lahore | 302 |
| Manila | 306 |
| Mexico City | 310 |
| Mumbai | 314 |
| Seoul | 318 |
| Shanghai | 322 |
| Tehran | 326 |

## A4.1 Beijing

Population size 2015:                     **16.2 million inhabitants**
Avg. Population density 2015:             **5200 / km²**
Estimated urban flood extent 2015:        **132.5 km² (rank 4)**
Estimated urban flood extent 2060:        **341.4 km² (rank 4)**

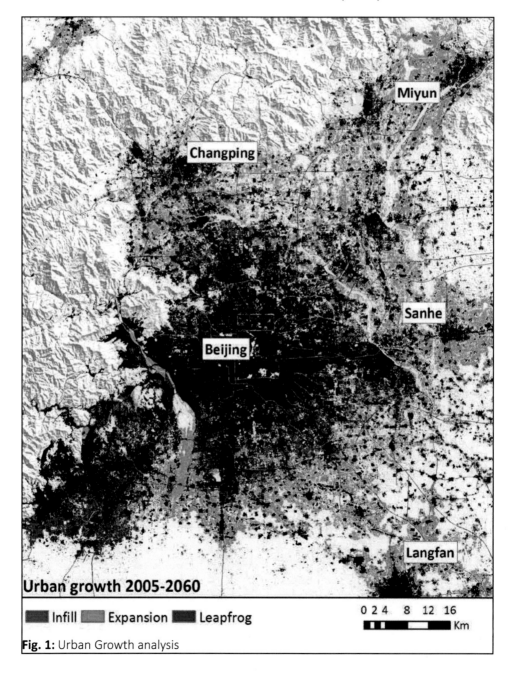

**Fig. 1:** Urban Growth analysis

Growth Rate Urban Flood Extent 2015-2060:      **2.60**
Growth Rate Safe Urban Extent 2015-2060:       **1.63**
Proportion Flooded Urban Extent in 2060:       **0.51%**

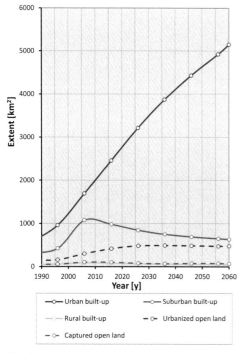

**Fig. 2:** Urban landscape analysis 1990- 2060

**Fig. 3:** Projected Urban Flood extent

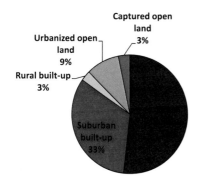

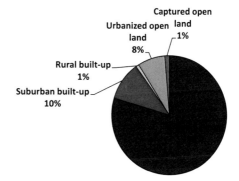

**Fig. 4:** PComposition of Urban Footprint for 2005(left) and 2060 (right)

# Beijing

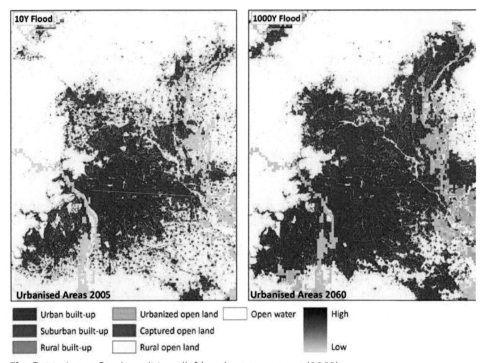

**Fig. 5:** Moderate flood conditions (left) and extreme event (2060)

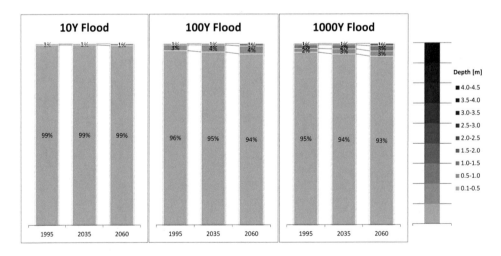

**Fig. 6:** Flood depth distribution

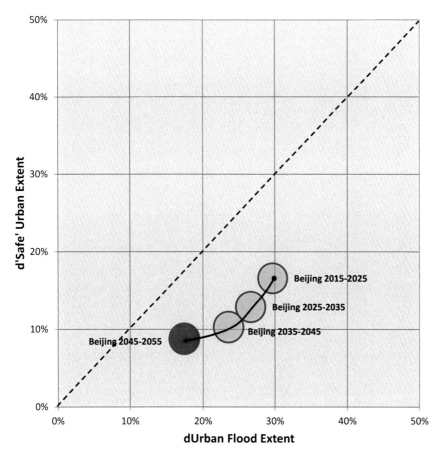

**Fig. 7:** Urban Growth Characterisation for 10Y intervals between 2015-2055

## A4.2 Cairo

Population size 2015:                       **15.9 million inhabitants**
Avg. Population density 2015:               **9000 / km²**
Estimated urban flood extent 2015:          **49.4 km² (rank 11)**
Estimated urban flood extent 2060:          **106.4 km² (rank 12)**

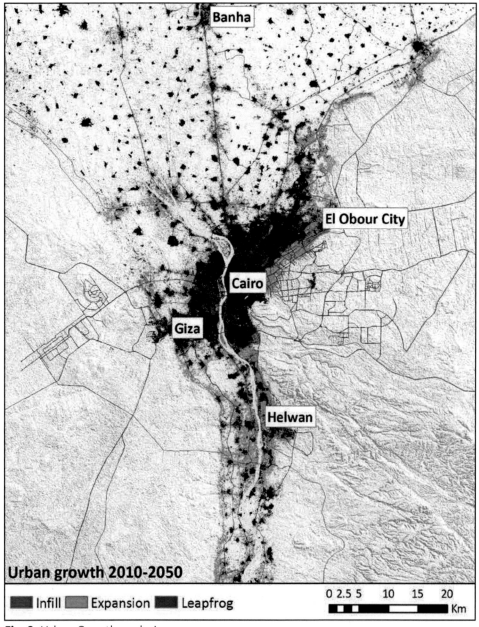

**Fig. 8:** Urban Growth analysis

Growth Rate Urban Flood Extent 2015-2060: **2.15**
Growth Rate Safe Urban Extent 2015-2060: **1.87**
Proportion Flooded Urban Extent in 2060: **9.54%**

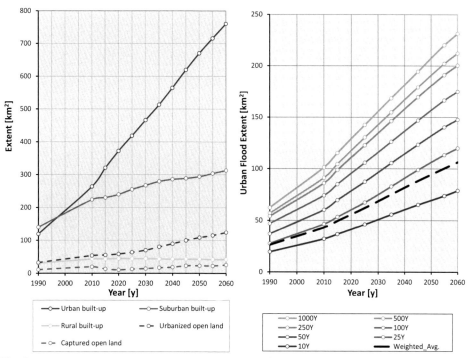

**Fig. 9:** Urban landscape analysis 1990- 2060      **Fig. 10:** Projected Urban Flood extent

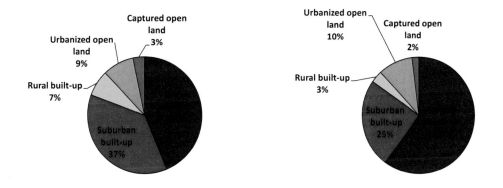

**Fig. 11:** PComposition of Urban Footprint for 2005(left) and 2060 (right)

# Cairo

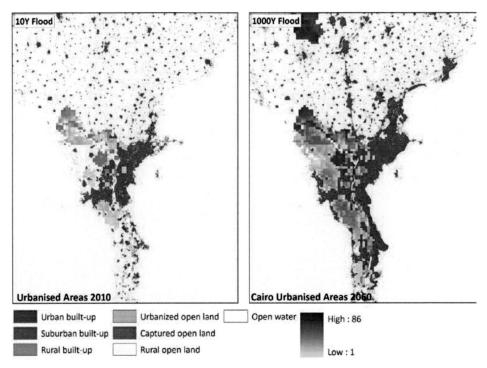

**Fig. 12:** Moderate flood conditions (left) and extreme event (2060)

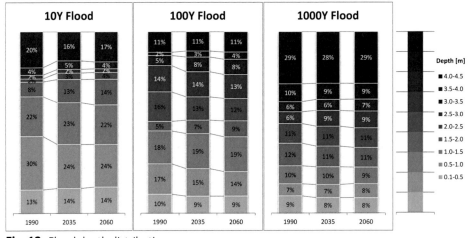

**Fig. 13:** Flood depth distribution

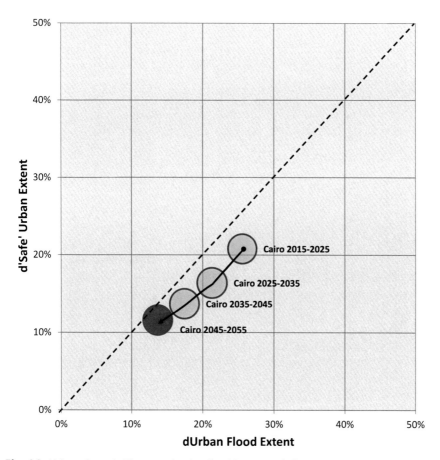

**Fig. 14:** Urban Growth Characterisation for 10Y intervals between 2015-2055

## A4.3 Calcutta

| | |
|---|---|
| Population size 2015: | **14.8 million inhabitants** |
| Avg. Population density 2015: | **12300 / km²** |
| Estimated urban flood extent 2015: | **198.2 km² (rank 3)** |
| Estimated urban flood extent 2060: | **408.6 km² (rank 2)** |

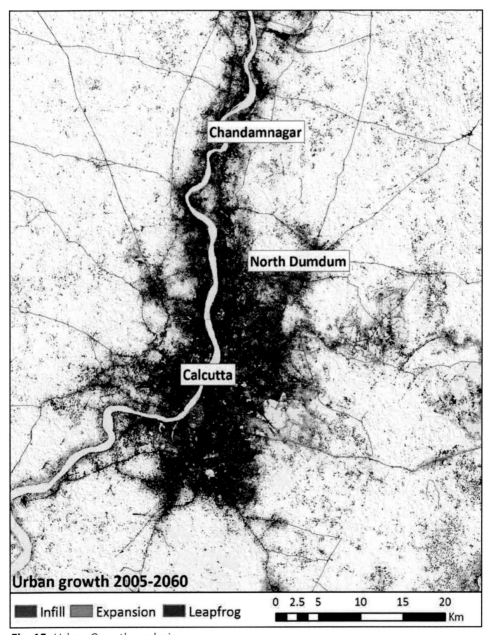

**Fig. 15:** Urban Growth analysis

Growth Rate Urban Flood Extent 2015-2060: **2.06**
Growth Rate Safe Urban Extent 2015-2060: **1.77**
Proportion Flooded Urban Extent in 2060: **34.67%**

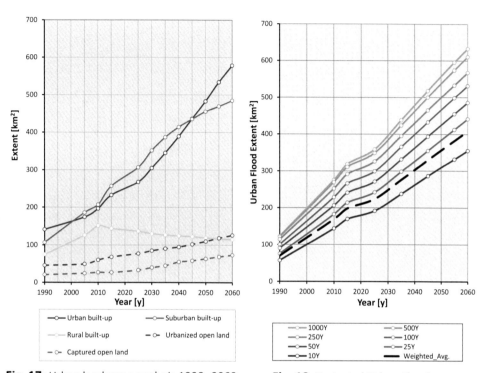

**Fig. 17:** Urban landscape analysis 1990- 2060    **Fig. 18:** Projected Urban Flood extent

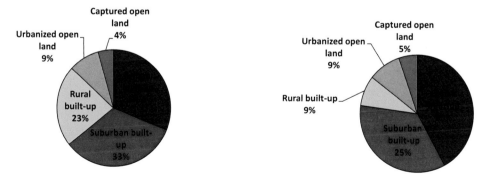

**Fig. 16:** PComposition of Urban Footprint for 2005(left) and 2060 (right)

# Calcutta

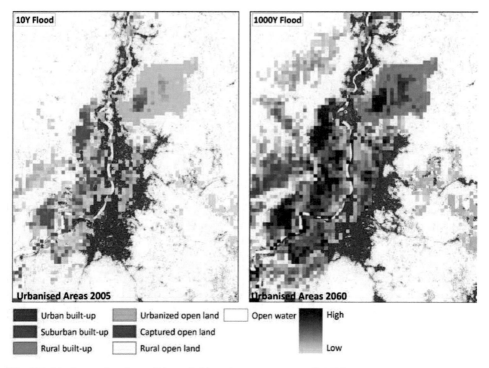

**Fig. 20:** Moderate flood conditions (left) and extreme event (2060)

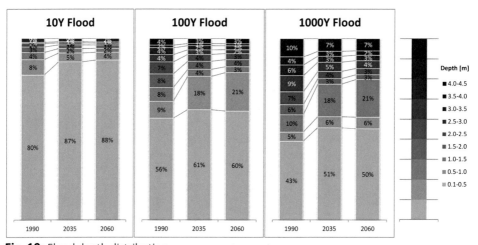

**Fig. 19:** Flood depth distribution

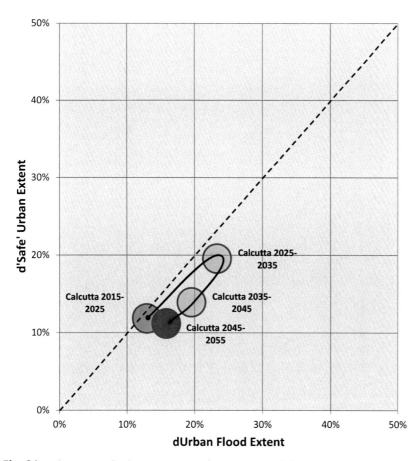

**Fig. 21:** Urban Growth Characterisation for 10Y intervals between 2015-2055

## A4.4 Delhi

Population size 2015:                        **25.7 million inhabitants**
Avg. Population density 2015:                **11900 / km²**
Estimated urban flood extent 2015:           **117.1 km² (rank 6)**
Estimated urban flood extent 2060:           **288.6 km² (rank 5)**

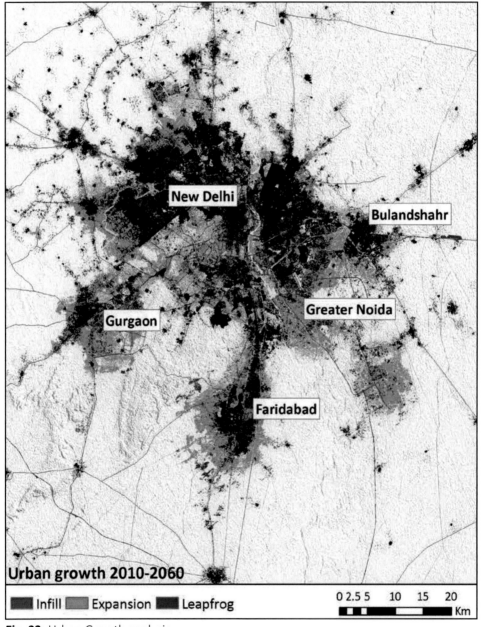

**Fig. 22:** Urban Growth analysis

Growth Rate Urban Flood Extent 2015-2060:     **2.46**
Growth Rate Safe Urban Extent 2015-2060:      **2.18**
Proportion Flooded Urban Extent in 2060:      **12.92%**

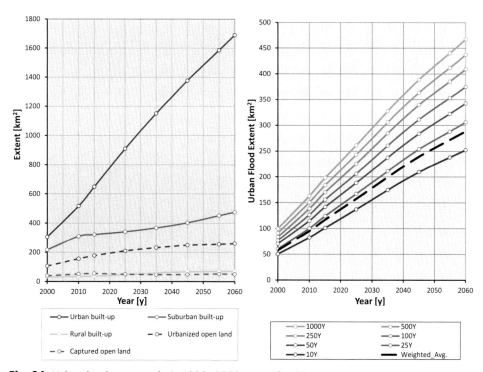

**Fig. 24:** Urban landscape analysis 1990- 2060      **Fig. 25:** Projected Urban Flood extent

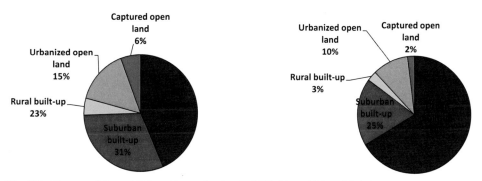

**Fig. 23:** PComposition of Urban Footprint for 2005(left) and 2060 (right)

# Delhi

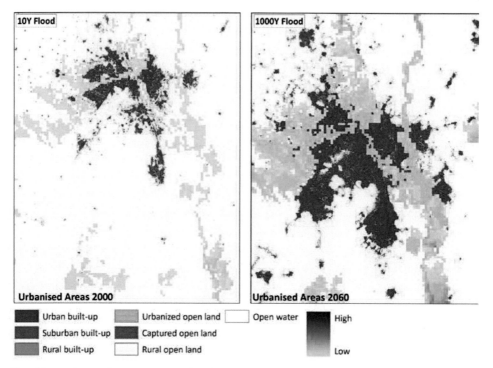

**Fig. 27:** Moderate flood conditions (left) and extreme event (2060)

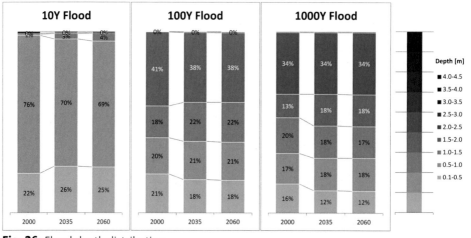

**Fig. 26:** Flood depth distribution

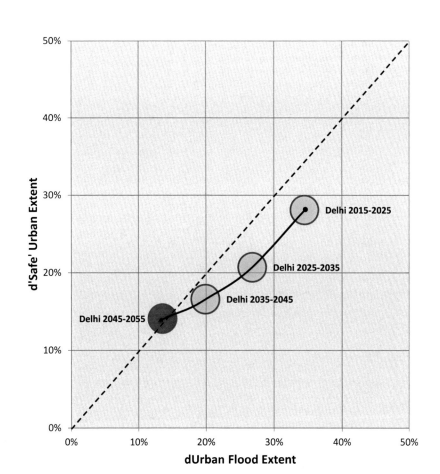

**Fig. 28:** Urban Growth Characterisation for 10Y intervals between 2015-2055

## A4.5 Dhaka

Population size 2015:                          **16.2 million inhabitants**
Avg. Population density 2015:                  **44100 / km²**
Estimated urban flood extent 2015:             **63.8 km² (rank 9)**
Estimated urban flood extent 2060:             **271.0 km² (rank 6)**

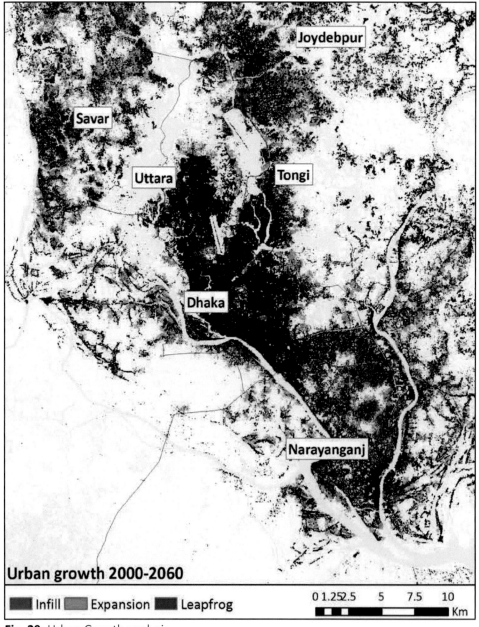

**Fig. 29:** Urban Growth analysis

Growth Rate Urban Flood Extent 2015-2060: **3.24**
Growth Rate Safe Urban Extent 2015-2060: **0.84**
Proportion Flooded Urban Extent in 2060: **49.74%**

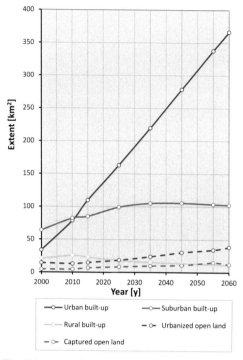

**Fig. 31:** Urban landscape analysis 1990- 2060

**Fig. 32:** Projected Urban Flood extent

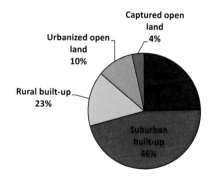

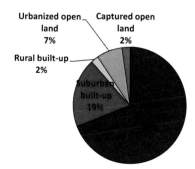

**Fig. 30:** PComposition of Urban Footprint for 2005(left) and 2060 (right)

# Dhaka

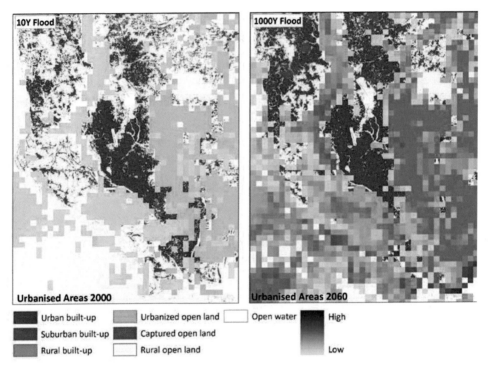

**Fig. 34:** Moderate flood conditions (left) and extreme event (2060)

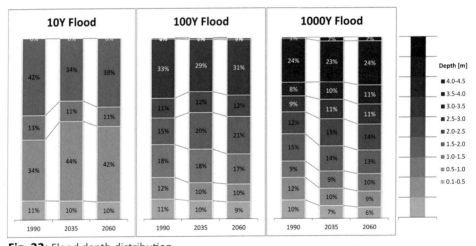

**Fig. 33:** Flood depth distribution

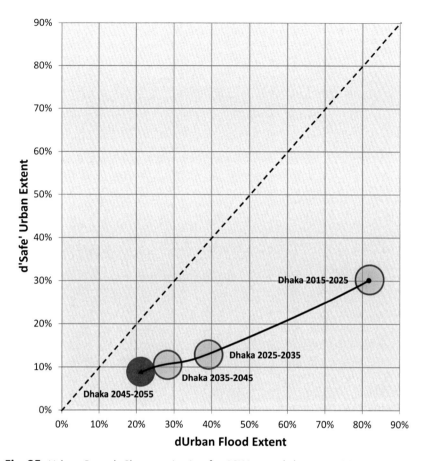

**Fig. 35:** Urban Growth Characterisation for 10Y intervals between 2015-2055

## A4.6 Guangzhou-Shenzen

| | |
|---|---|
| Population size 2015: | **31 million inhabitants** |
| Avg. Population density 2015: | **5729 / km²** |
| Estimated urban flood extent 2015: | **668.0 km² (rank 1)** |
| Estimated urban flood extent 2060: | **1417.4 km² (rank 1)** |

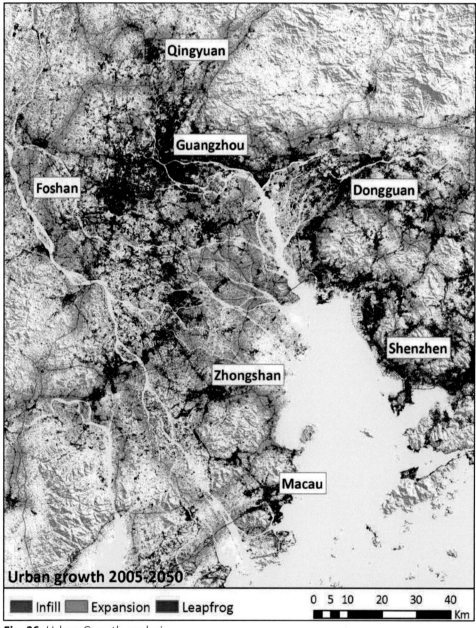

**Fig. 36:** Urban Growth analysis

Growth Rate Urban Flood Extent 2015-2060:    **2.10**
Growth Rate Safe Urban Extent 2015-2060:     **1.92**
Proportion Flooded Urban Extent in 2060:      **16.45%**

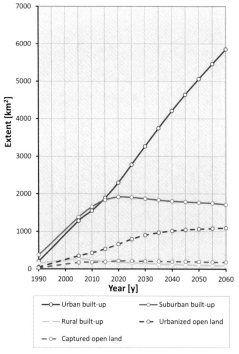

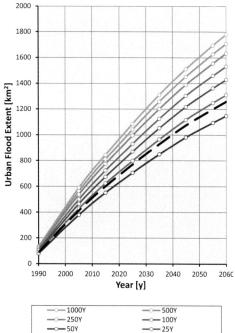

**Fig. 38:** Urban landscape analysis 1990- 2060

**Fig. 39:** Projected Urban Flood extent

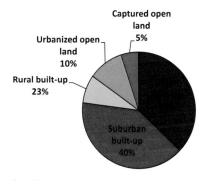

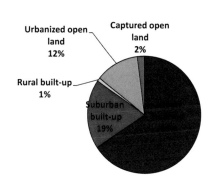

**Fig. 37:** PComposition of Urban Footprint for 2005(left) and 2060 (right)

# Guangzhou-Shenzen

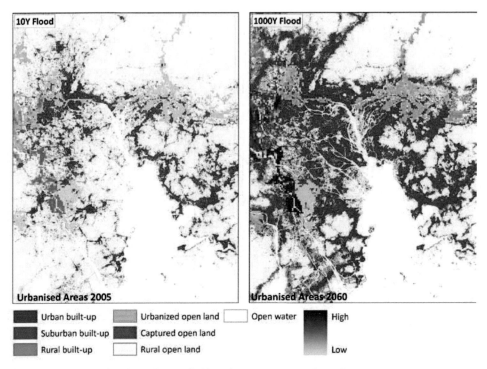

**Fig. 41:** Moderate flood conditions (left) and extreme event (2060)

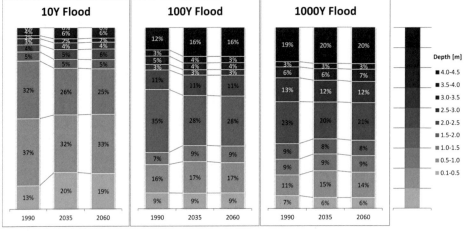

**Fig. 40:** Flood depth distribution

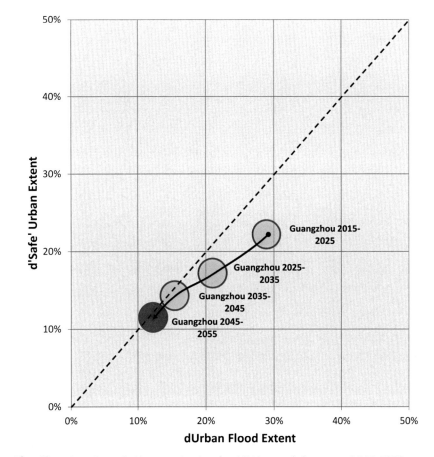

**Fig. 42:** Urban Growth Characterisation for 10Y intervals between 2015-2055

## A4.7 Ho Chi Minh City

| | |
|---|---|
| Population size 2015: | **10.1 million inhabitants** |
| Avg. Population density 2015: | **6500 / km²** |
| Estimated urban flood extent 2015: | **44.2 km² (rank 12)** |
| Estimated urban flood extent 2060: | **182.5 km² (rank 8)** |

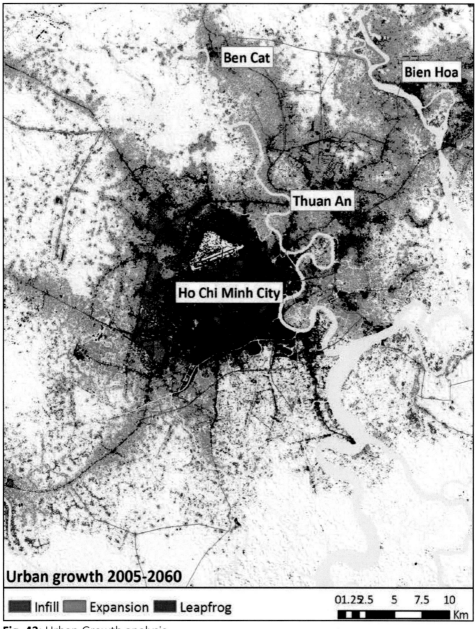

**Fig. 43:** Urban Growth analysis

Growth Rate Urban Flood Extent 2015-2060:     **4.13**
Growth Rate Safe Urban Extent 2015-2060:      **1.90**
Proportion Flooded Urban Extent in 2060:      **23.76%**

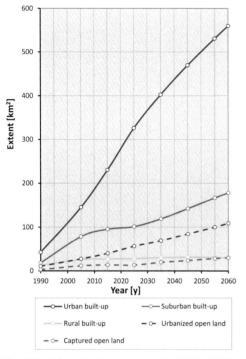

**Fig. 45:** Urban landscape analysis 1990- 2060

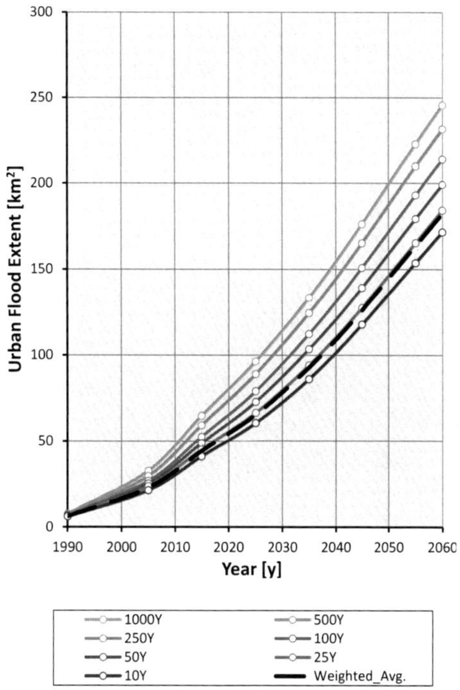

**Fig. 46:** Projected Urban Flood extent

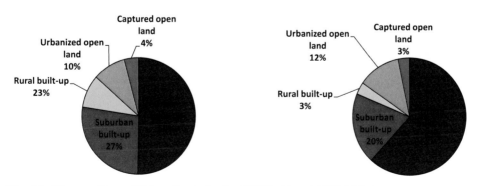

**Fig. 44:** PComposition of Urban Footprint for 2005(left) and 2060 (right)

# Ho Chi Minh City

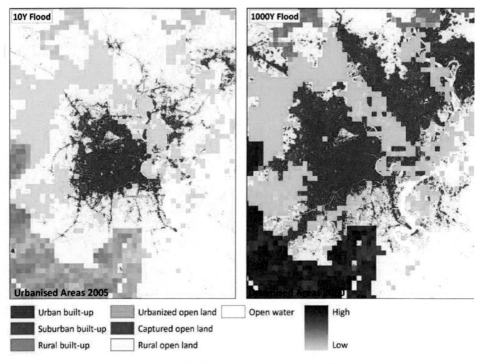

**Fig. 48:** Moderate flood conditions (left) and extreme event (2060)

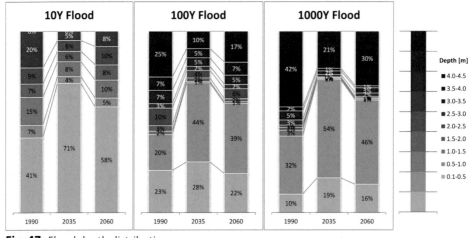

**Fig. 47:** Flood depth distribution

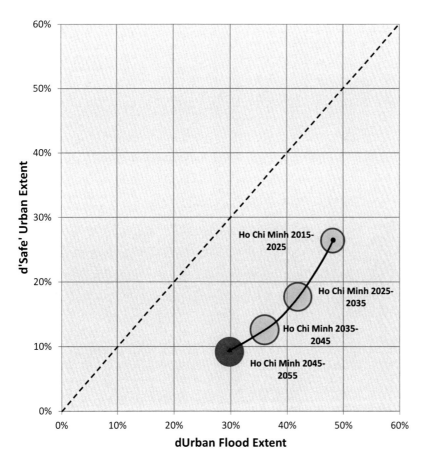

**Fig. 49:** Urban Growth Characterisation for 10Y intervals between 2015-2055

# A4.8 Istanbul

Population size 2015:                     **13.5 million inhabitants**
Avg. Population density 2015:             **9900 / km²**
Estimated urban flood extent 2015:        **1.0 km² (rank 18)**
Estimated urban flood extent 2060:        **4.6 km² (rank 18)**

**Fig. 50:** Urban Growth analysis

Growth Rate Urban Flood Extent 2015-2060:     **4.66**
Growth Rate Safe Urban Extent 2015-2060:      **1.70**
Proportion Flooded Urban Extent in 2060:      **0.31%**

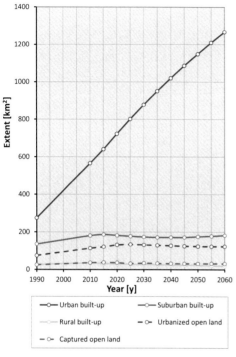

**Fig. 52:** Urban landscape analysis 1990- 2060

**Fig. 53:** Projected Urban Flood extent

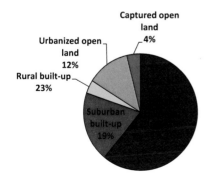

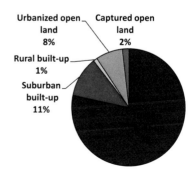

**Fig. 51:** PComposition of Urban Footprint for 2005(left) and 2060 (right)

# Istanbul

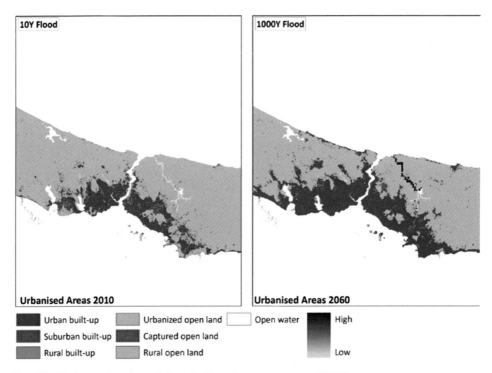

**Fig. 55:** Moderate flood conditions (left) and extreme event (2060)

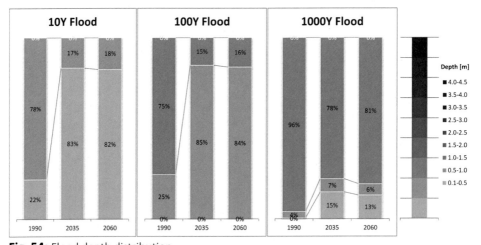

**Fig. 54:** Flood depth distribution

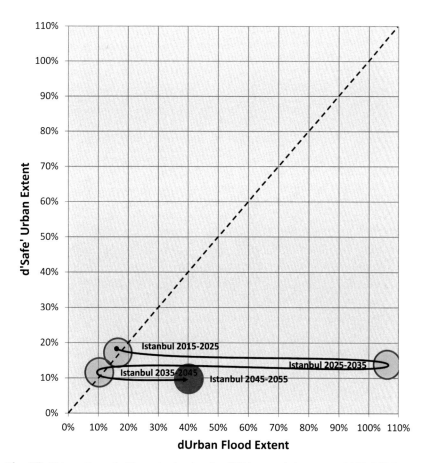

**Fig. 56:** Urban Growth Characterisation for 10Y intervals between 2015-2055

# A4.9 Jakarta

| | |
|---|---|
| Population size 2015: | **31.3 million inhabitants** |
| Avg. Population density 2015: | **9700 / km²** |
| Estimated urban flood extent 2015: | **70.0 km² (rank 8)** |
| Estimated urban flood extent 2060: | **216.1 km² (rank 7)** |

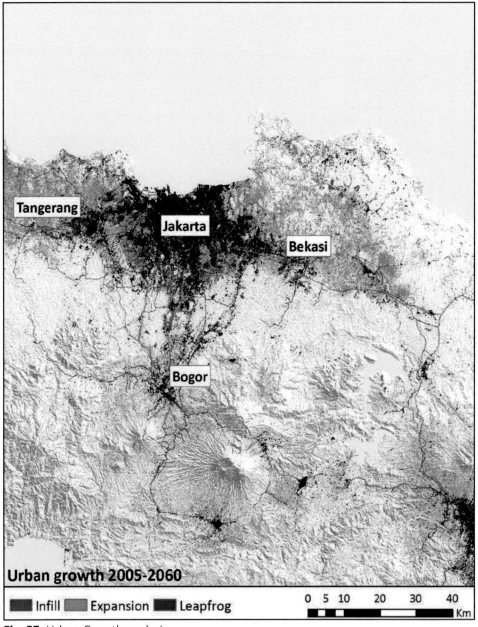

**Fig. 57:** Urban Growth analysis

Growth Rate Urban Flood Extent 2015-2060: **3.09**
Growth Rate Safe Urban Extent 2015-2060: **1.96**
Proportion Flooded Urban Extent in 2060: **8.28%**

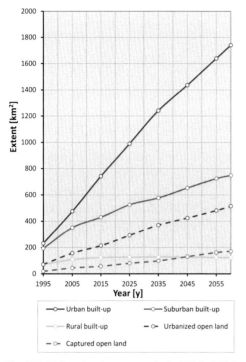

**Fig. 59:** Urban landscape analysis 1990- 2060

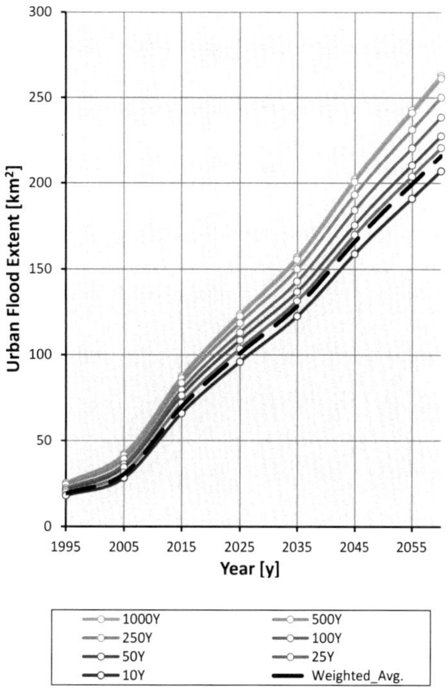

**Fig. 60:** Projected Urban Flood extent

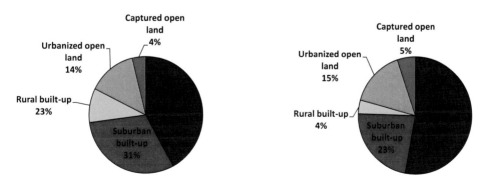

**Fig. 58:** PComposition of Urban Footprint for 2005(left) and 2060 (right)

# Jakarta

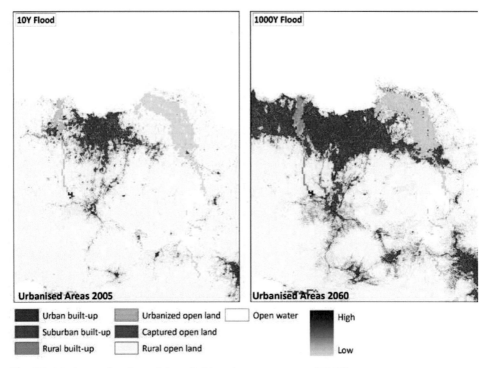

**Fig. 62:** Moderate flood conditions (left) and extreme event (2060)

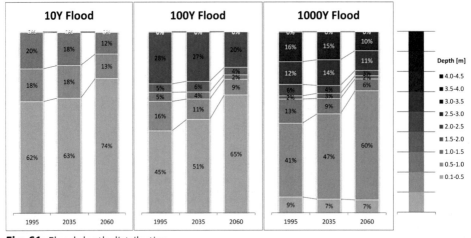

**Fig. 61:** Flood depth distribution

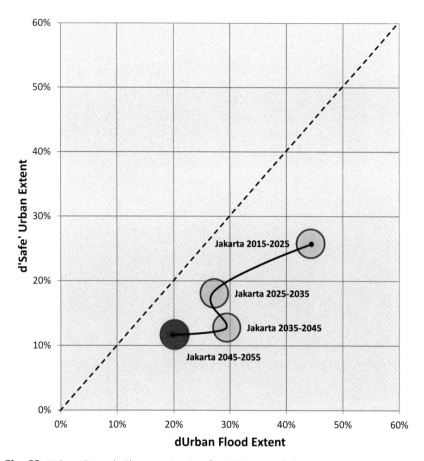

**Fig. 63:** Urban Growth Characterisation for 10Y intervals between 2015-2055

## A4.10  Karachi

Population size 2015:                    **22.8 million inhabitants**
Avg. Population density 2015:            **24100 / km²**
Estimated urban flood extent 2015:      **7.3 km² (rank 16)**
Estimated urban flood extent 2060:      **19.0 km² (rank 16)**

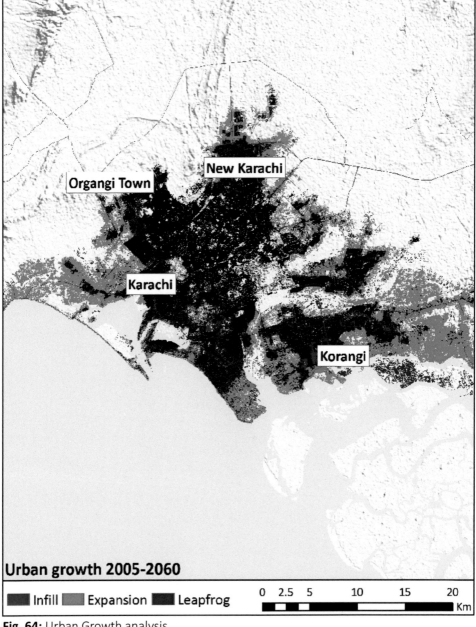

**Fig. 64:** Urban Growth analysis

Growth Rate Urban Flood Extent 2015-2060: **2.61**
Growth Rate Safe Urban Extent 2015-2060: **2.21**
Proportion Flooded Urban Extent in 2060: **0.85%**

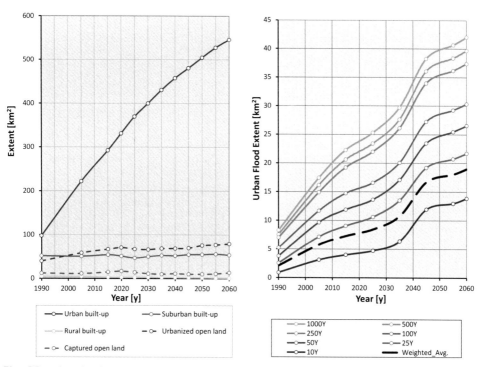

**Fig. 66:** Urban landscape analysis 1990- 2060          **Fig. 67:** Projected Urban Flood extent

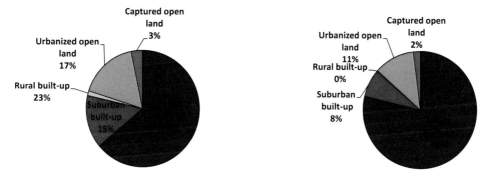

**Fig. 65:** PComposition of Urban Footprint for 2005(left) and 2060 (right)

# Karachi

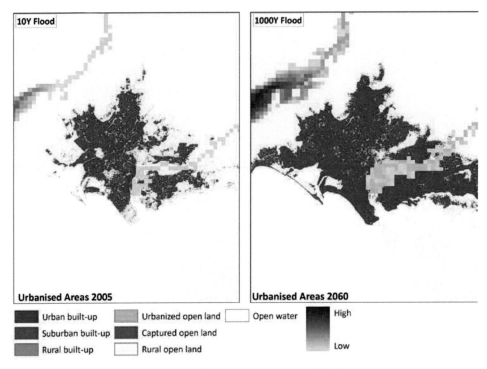

**Fig. 69:** Moderate flood conditions (left) and extreme event (2060)

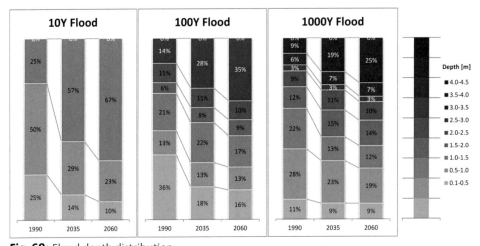

**Fig. 68:** Flood depth distribution

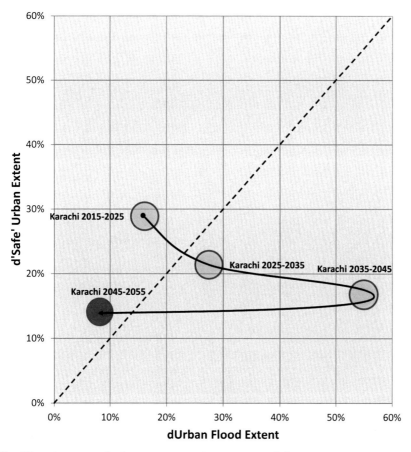

**Fig. 70:** Urban Growth Characterisation for 10Y intervals between 2015-2055

## A4.11 Lagos

Population size 2015:                    **12.8 million inhabitants**
Avg. Population density 2015:            **9000 / km²**
Estimated urban flood extent 2015:       **75.5 km² (rank 7)**
Estimated urban flood extent 2060:       **177.6 km² (rank 10)**

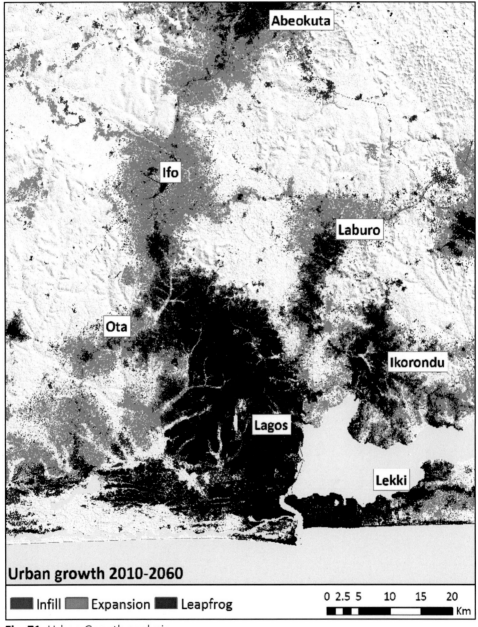

**Fig. 71:** Urban Growth analysis

Growth Rate Urban Flood Extent 2015-2060: **2.35**
Growth Rate Safe Urban Extent 2015-2060: **2.41**
Proportion Flooded Urban Extent in 2060: **7.38%**

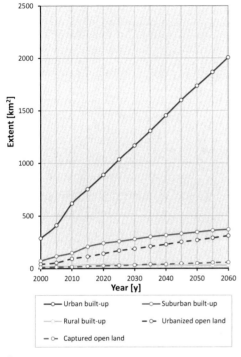

**Fig. 73:** Urban landscape analysis 1990- 2060

**Fig. 74:** Projected Urban Flood extent

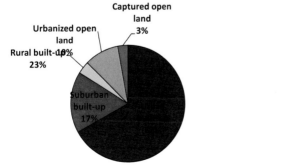

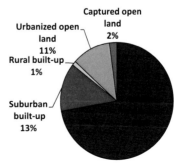

**Fig. 72:** PComposition of Urban Footprint for 2005(left) and 2060 (right)

# Lagos

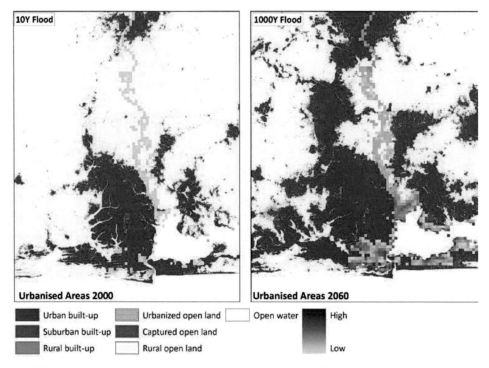

**Fig. 76:** Moderate flood conditions (left) and extreme event (2060)

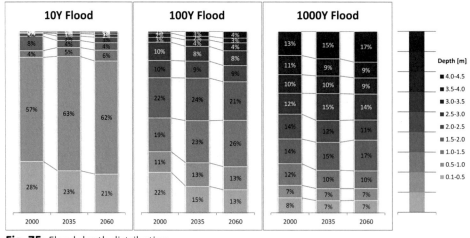

**Fig. 75:** Flood depth distribution

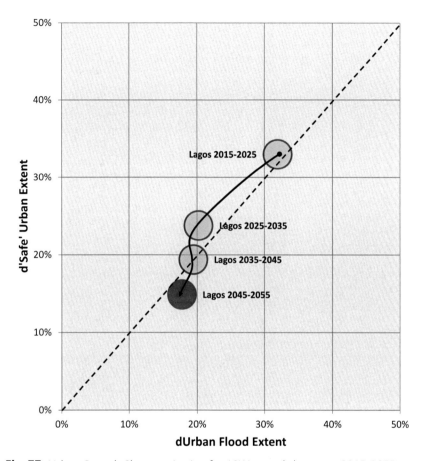

**Fig. 77:** Urban Growth Characterisation for 10Y intervals between 2015-2055

## A4.12 Lahore

| | |
|---|---|
| Population size 2015: | **10.4 million inhabitants** |
| Avg. Population density 2015: | **13100 / km²** |
| Estimated urban flood extent 2015: | **12.7 km² (rank 15)** |
| Estimated urban flood extent 2060: | **51.7 km² (rank 15)** |

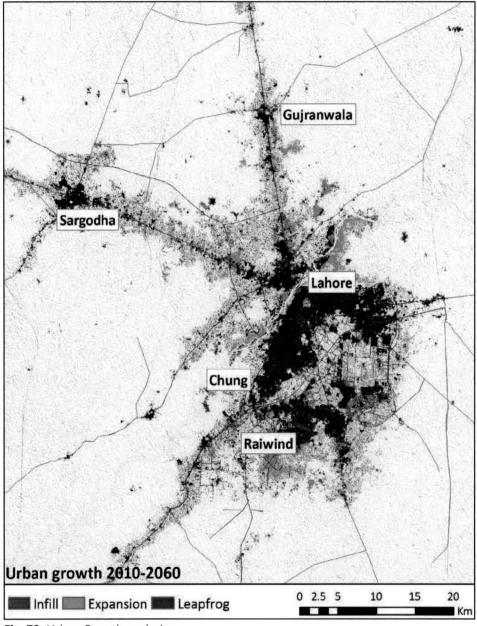

**Fig. 78:** Urban Growth analysis

Growth Rate Urban Flood Extent 2015-2060:     **4.08**
Growth Rate Safe Urban Extent 2015-2060:      **1.96**
Proportion Flooded Urban Extent in 2060:       **9.04%**

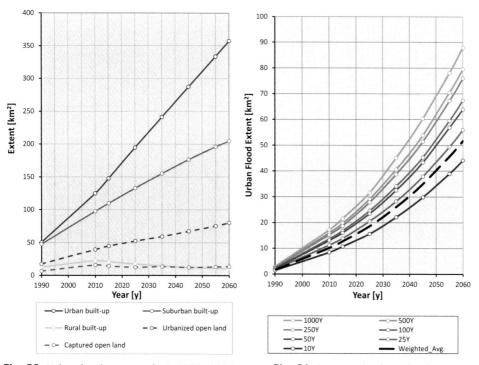

**Fig. 80:** Urban landscape analysis 1990- 2060

**Fig. 81:** Projected Urban Flood extent

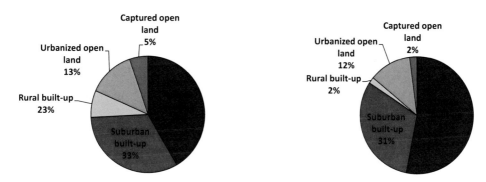

**Fig. 79:** PComposition of Urban Footprint for 2005(left) and 2060 (right)

# Lahore

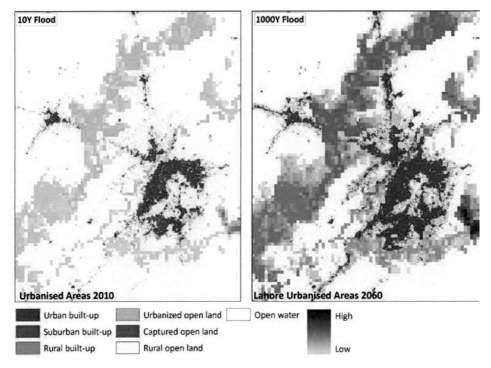

**Fig. 83:** Moderate flood conditions (left) and extreme event (2060)

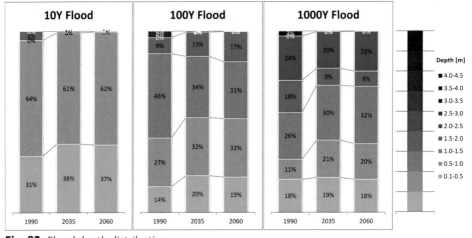

**Fig. 82:** Flood depth distribution

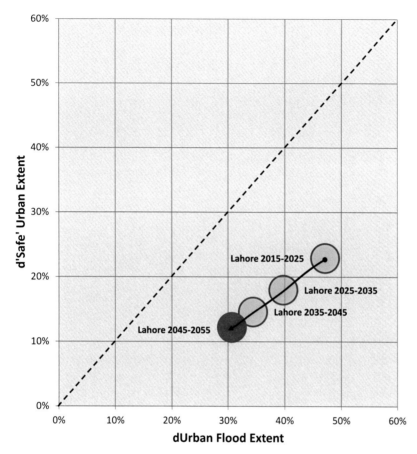

**Fig. 84:** Urban Growth Characterisation for 10Y intervals between 2015-2055

# A4.13 Manila

Population size 2015:                    **22.9 million inhabitants**
Avg. Population density 2015:            **14100 / km²**
Estimated urban flood extent 2015:      **43.9 km² (rank 13)**
Estimated urban flood extent 2060:      **107.7 km² (rank 11)**

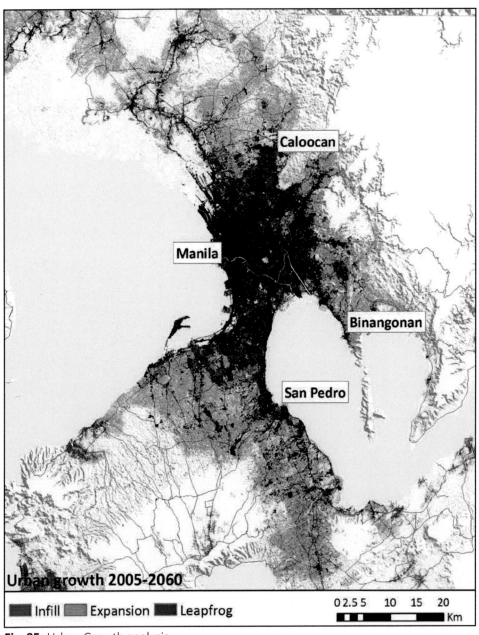

**Fig. 85:** Urban Growth analysis

Growth Rate Urban Flood Extent 2015-2060: **2.46**
Growth Rate Safe Urban Extent 2015-2060: **2.19**
Proportion Flooded Urban Extent in 2060: **5.70%**

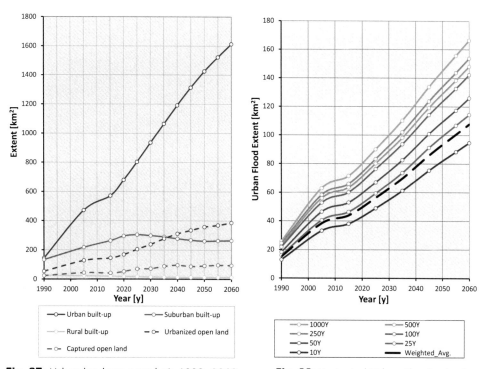

**Fig. 87:** Urban landscape analysis 1990- 2060

**Fig. 88:** Projected Urban Flood extent

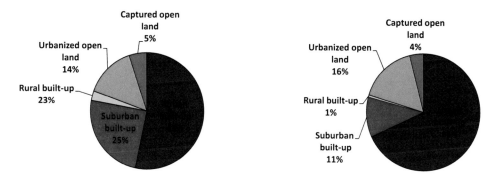

**Fig. 86:** PComposition of Urban Footprint for 2005(left) and 2060 (right)

# Manila

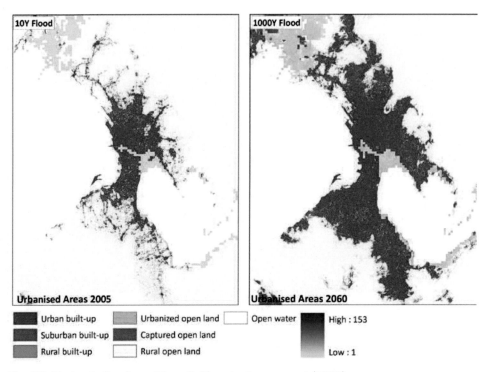

**Fig. 90:** Moderate flood conditions (left) and extreme event (2060)

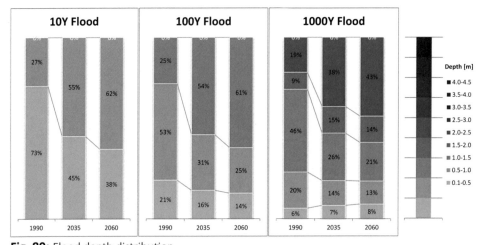

**Fig. 89:** Flood depth distribution

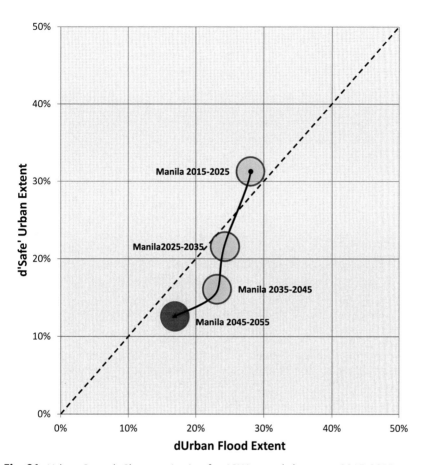

**Fig. 91:** Urban Growth Characterisation for 10Y intervals between 2015-2055

## A4.14  Mexico City

Population size 2015:                        **20.2 million inhabitants**
Avg. Population density 2015:                **9800 / km²**
Estimated urban flood extent 2015:           **128.1 km² (rank 5)**
Estimated urban flood extent 2060:           **179.9 km² (rank 9)**

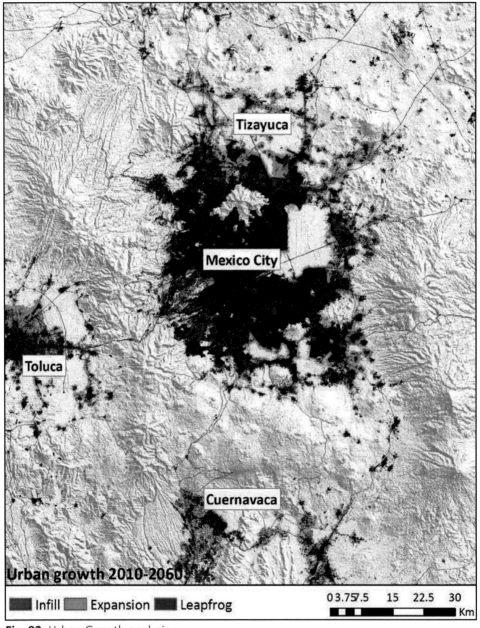

**Fig. 92:** Urban Growth analysis

Growth Rate Urban Flood Extent 2015-2060: **1.40**
Growth Rate Safe Urban Extent 2015-2060: **1.59**
Proportion Flooded Urban Extent in 2060: **5.66%**

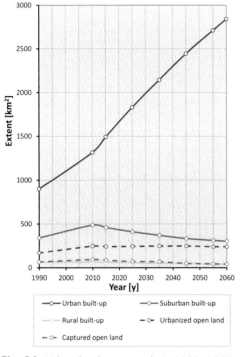

**Fig. 94:** Urban landscape analysis 1990- 2060

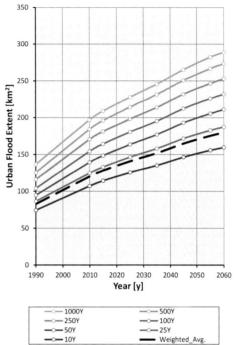

**Fig. 95:** Projected Urban Flood extent

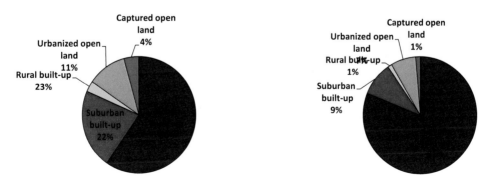

**Fig. 93:** PComposition of Urban Footprint for 2005(left) and 2060 (right)

# Mexico City

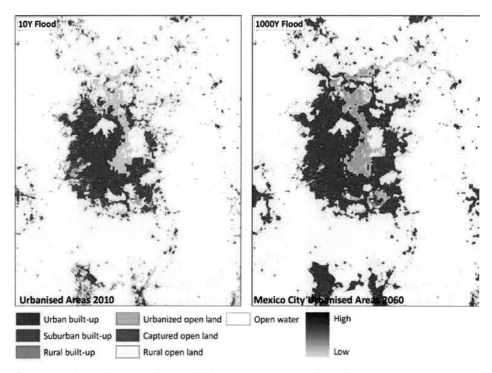

**Fig. 97:** Moderate flood conditions (left) and extreme event (2060)

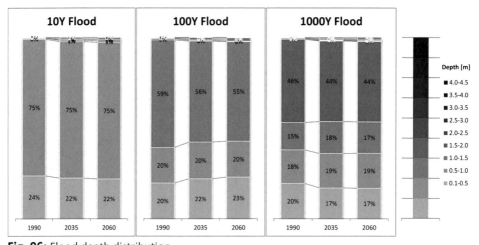

**Fig. 96:** Flood depth distribution

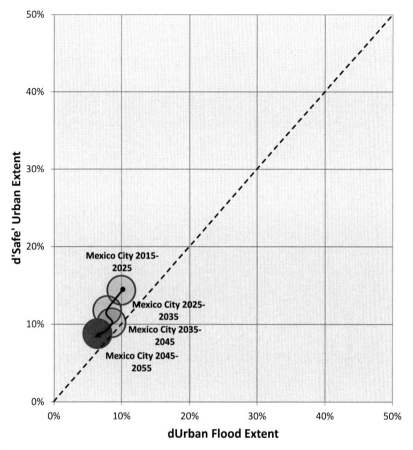

**Fig. 98:** Urban Growth Characterisation for 10Y intervals between 2015-2055

# A4.15 Mumbai

| | |
|---|---|
| Population size 2015: | **22.9 million inhabitants** |
| Avg. Population density 2015: | **26000 / km²** |
| Estimated urban flood extent 2015: | **23.0 km² (rank 14)** |
| Estimated urban flood extent 2060: | **56.1 km² (rank 14)** |

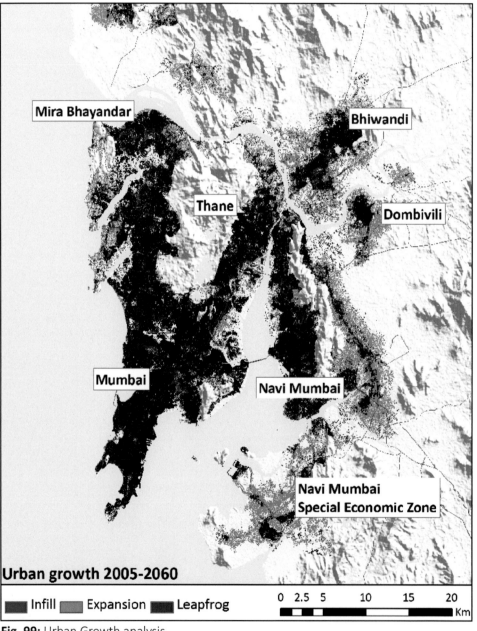

**Fig. 99:** Urban Growth analysis

Growth Rate Urban Flood Extent 2015-2060:     **2.44**
Growth Rate Safe Urban Extent 2015-2060:      **1.58**
Proportion Flooded Urban Extent in 2060:      **6.74%**

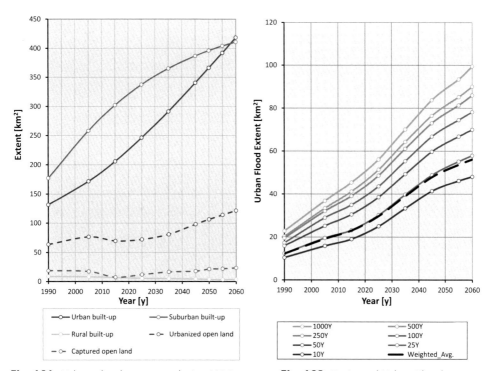

**Fig. 101:** Urban  landscape  analysis  1990 -

**Fig. 102:** Projected Urban Flood extent

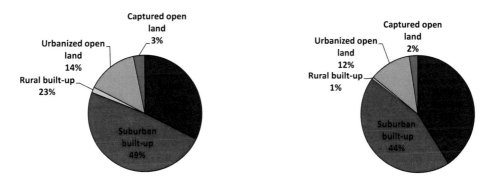

**Fig. 100:** PComposition of Urban Footprint for 2005(left) and 2060 (right)

# Mumbai

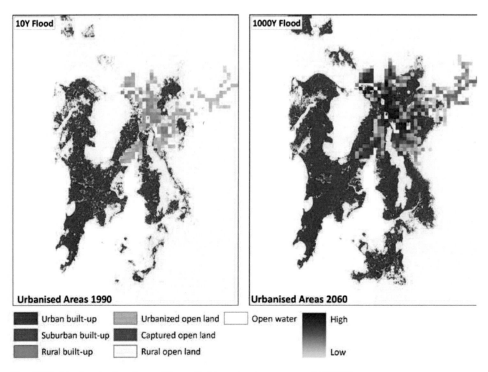

**Fig. 104:** Moderate flood conditions (left) and extreme event (2060)

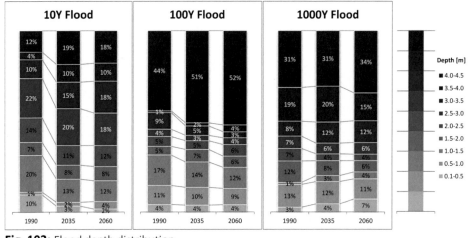

**Fig. 103:** Flood depth distribution

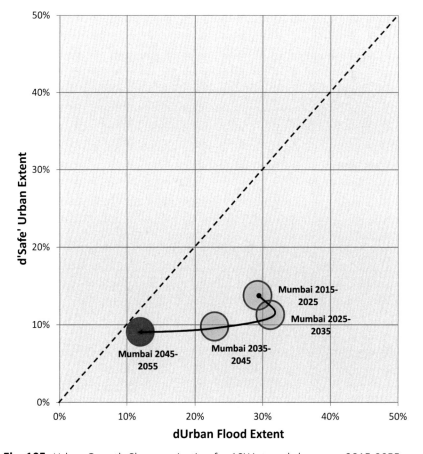

**Fig. 105:** Urban Growth Characterisation for 10Y intervals between 2015-2055

# A4.16 Seoul

| | |
|---|---|
| Population size 2015: | **23.6 million inhabitants** |
| Avg. Population density 2015: | **9100 / km²** |
| Estimated urban flood extent 2015: | **55.1 km² (rank 10)** |
| Estimated urban flood extent 2060: | **80.5 km² (rank 13)** |

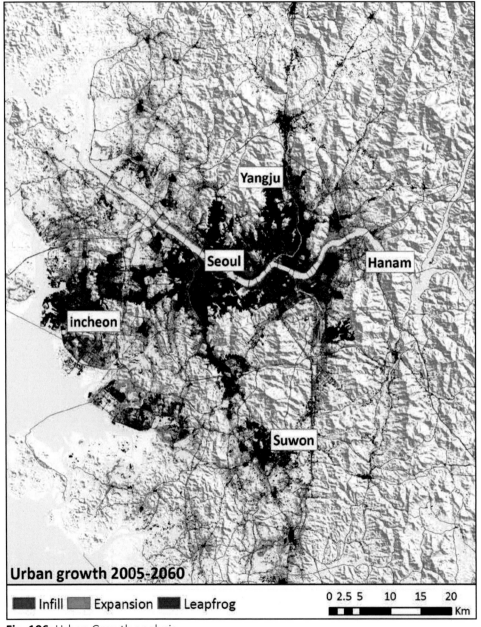

**Fig. 106:** Urban Growth analysis

Growth Rate Urban Flood Extent 2015-2060: **1.45**
Growth Rate Safe Urban Extent 2015-2060: **1.71**
Proportion Flooded Urban Extent in 2060: **4.16%**

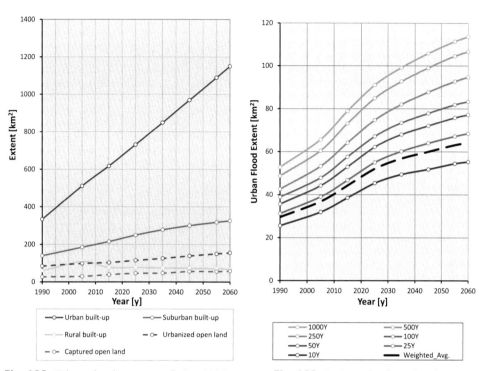

**Fig. 108:** Urban landscape analysis 1990 -

**Fig. 109:** Projected Urban Flood extent

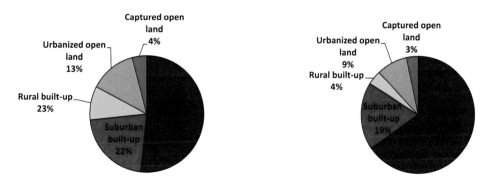

**Fig. 107:** PComposition of Urban Footprint for 2005(left) and 2060 (right)

# Seoul

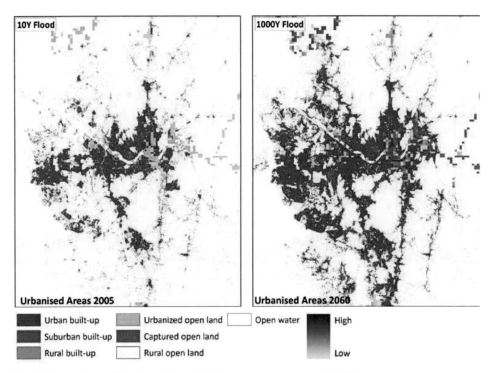

**Fig. 111:** Moderate flood conditions (left) and extreme event (2060)

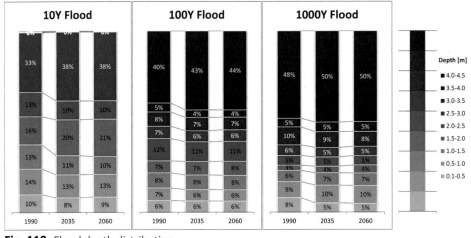

**Fig. 110:** Flood depth distribution

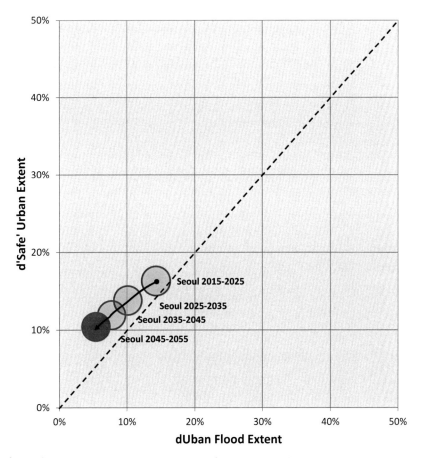

**Fig. 112:** Urban Growth Characterisation for 10Y intervals between 2015-2055

# A4.17  Shanghai

Population size 2015:                    **22.7 million inhabitants**
Avg. Population density 2015:            **5800 / km²**
Estimated urban flood extent 2015:       **219.2 km² (rank 2)**
Estimated urban flood extent 2060:       **347.3 km² (rank 3)**

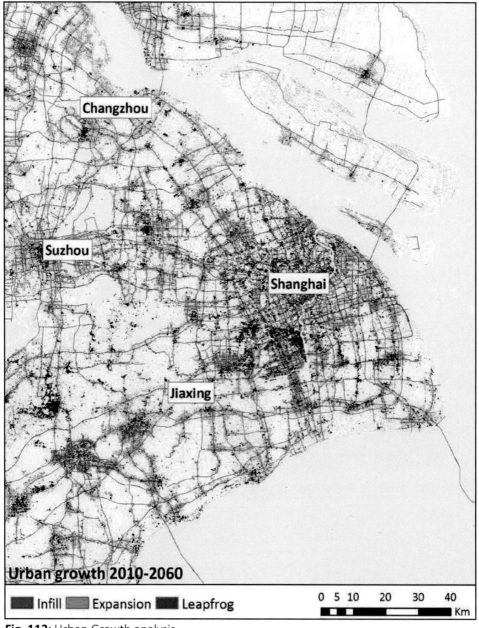

**Fig. 113:** Urban Growth analysis

Growth Rate Urban Flood Extent 2015-2060:    **1.59**
Growth Rate Safe Urban Extent 2015-2060:    **1.89**
Proportion Flooded Urban Extent in 2060:    **12.66%**

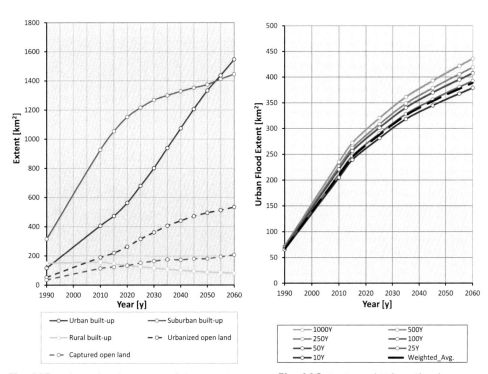

**Fig. 115:** Urban landscape analysis 1990 -

**Fig. 116:** Projected Urban Flood extent

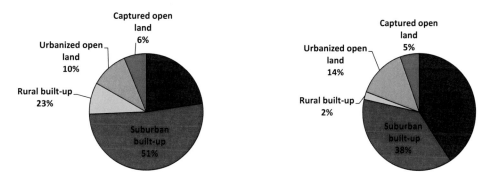

**Fig. 114:** PComposition of Urban Footprint for 2005(left) and 2060 (right)

# Shanghai

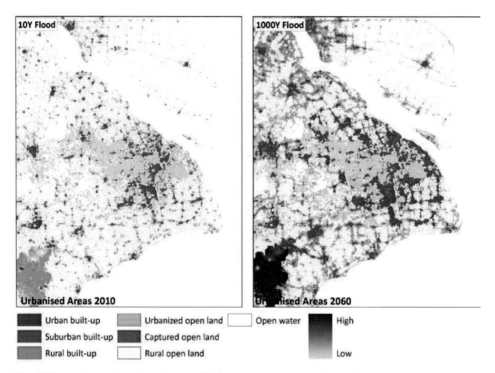

**Fig. 118:** Moderate flood conditions (left) and extreme event (2060)

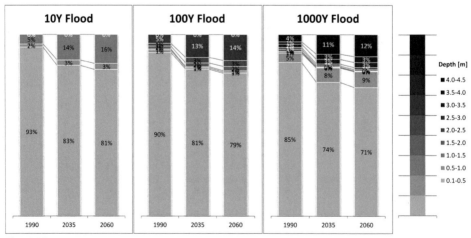

**Fig. 117:** Flood depth distribution

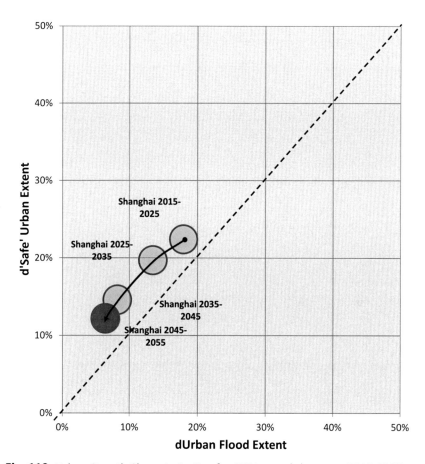

**Fig. 119:** Urban Growth Characterisation for 10Y intervals between 2015-2055

# A4.18 Tehran

Population size 2015:                           **13.7 million inhabitants**
Avg. Population density 2015:         **8400 / km²**
Estimated urban flood extent 2015:   **5.2 km² (rank 17)**
Estimated urban flood extent 2060:   **11.3 km² (rank 17)**

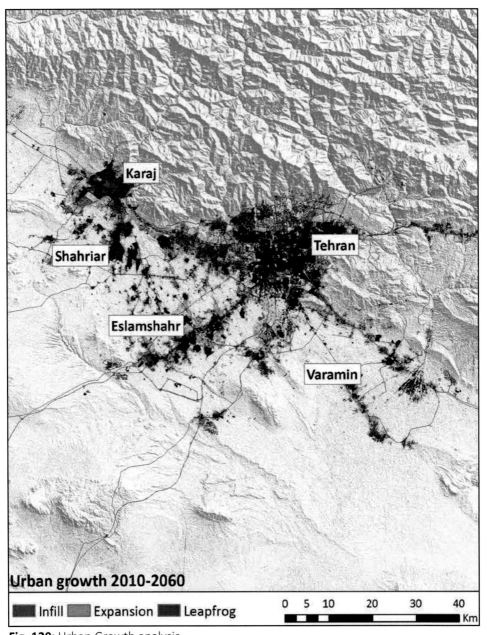

**Fig. 120:** Urban Growth analysis

Growth Rate Urban Flood Extent 2015-2060:   **2.18**
Growth Rate Safe Urban Extent 2015-2060:    **1.78**
Proportion Flooded Urban Extent in 2060:    **0.90%**

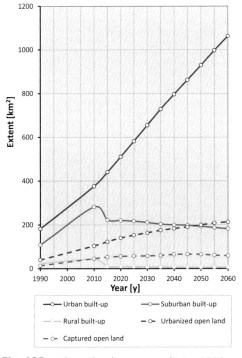

**Fig. 122:** Urban  landscape  analysis  1990 -

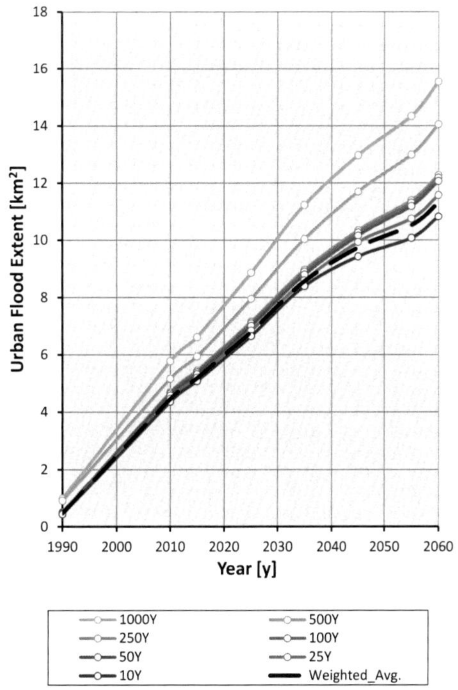

**Fig. 123:** Projected Urban Flood extent

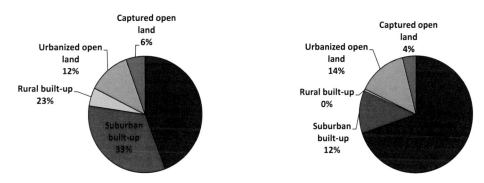

**Fig. 121:** PComposition of Urban Footprint for 2005(left) and 2060 (right)

# Tehran

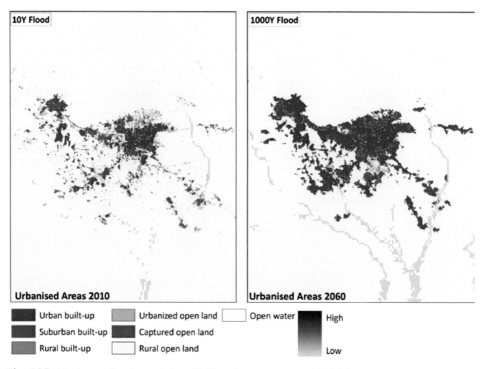

**Fig. 125:** Moderate flood conditions (left) and extreme event (2060)

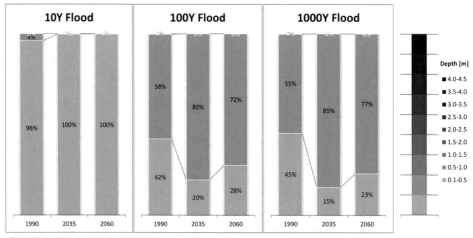

**Fig. 124:** Flood depth distribution

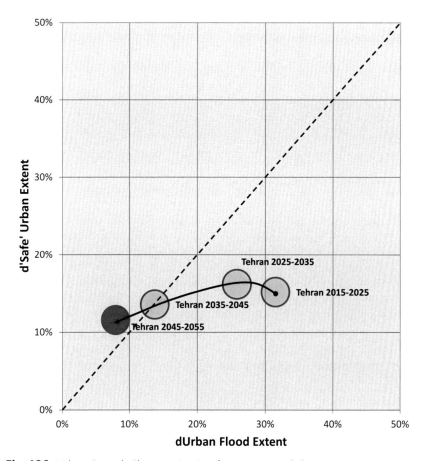

**Fig. 126:** Urban Growth Characterisation for 10Y intervals between 2015-2055

# A5 Urban growth ratio

The ratio $(GR)$ between the growth of the weighted mean urban flood extent as a function of urban growth and the difference between the 2010 urban flood extent associated to a 10Y and a 1000Y event, provides an indication if cities are more sensitive to urban growth or to shifting flood frequencies.

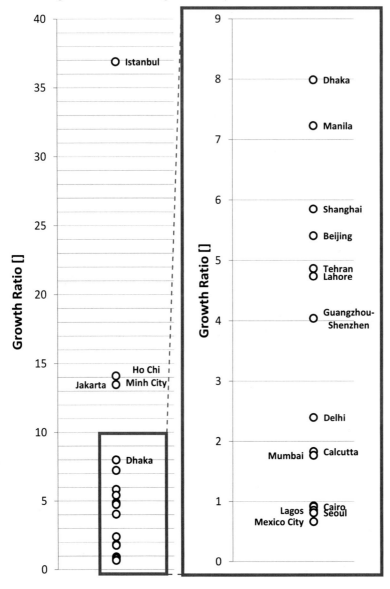

**Fig. 127:** Growth ratios for the highest scoring cites (left) and remaining cities (right)

# Appendix B: Pluvial flooding

# B1 Qualification of macro-, meso- and microscale indicators

**Table B1:** Ranges for qualifations for the respective indicators

| | | Threshold values [change 2015-2060] | | | |
|---|---|---|---|---|---|
| | | Average ISR | FD | Open Land Fraction | Disproportionate growth built-up areas |
| | Very Positive | <-20% | >3% | >20% | >20% |
| | Positive | <-10% | 1%-3% | 10%-20% | 10%-20% |
| | Neutral | -10%-10% | -1%-1% | -10%-10% | -10%-10% |
| | Negative | 10%-20% | -1%-3% | -10%--20% | -10%-20% |
| | Very Negative | >20% | <3% | <-20% | <-20% |

# B2 Macroscale indicators

**Table B2:** Estimated average impervious surface ratio for the urban footprint..

| | Avg. ISR | | | | | | |
|---|---|---|---|---|---|---|---|
| | 2005 | 2015 | 2025 | 2035 | 2045 | 2055 | 2060 |
| Beijing | 0.55 | 0.59 | 0.63 | 0.65 | 0.68 | 0.69 | 0.70 |
| Cairo | 0.52 | 0.53 | 0.54 | 0.55 | 0.57 | 0.58 | 0.58 |
| Calcutta | 0.44 | 0.42 | 0.43 | 0.44 | 0.46 | 0.47 | 0.48 |
| Dehli | 0.50 | 0.51 | 0.53 | 0.55 | 0.56 | 0.57 | 0.57 |
| Dhaka | 0.54 | 0.57 | 0.650 | 0.64 | 0.67 | 0.74 | 0.71 |
| Guangzhou-Shenzhen | 0.43 | 0.44 | 0.47 | 0.50 | 0.53 | 0.56 | 0.57 |
| Ho Chi Minh City | 0.49 | 0.55 | 0.58 | 0.60 | 0.60 | 0.59 | 0.59 |
| Istanbul | 0.54 | 0.57 | 0.61 | 0.64 | 0.66 | 0.68 | 0.69 |
| Jakarta | 0.47 | 0.48 | 0.49 | 0.51 | 0.52 | 0.53 | 0.53 |
| Karachi | 0.58 | 0.62 | 0.67 | 0.68 | 0.70 | 0.72 | 0.72 |
| Lagos | 0.57 | 0.55 | 0.57 | 0.59 | 0.650 | 0.61 | 0.62 |
| Lahore | 0.44 | 0.45 | 0.46 | 0.47 | 0.48 | 0.49 | 0.50 |
| Manila | 0.52 | 0.51 | 0.56 | 0.60 | 0.64 | 0.66 | 0.66 |
| Mexico City | 0.51 | 0.54 | 0.58 | 0.61 | 0.64 | 0.67 | 0.68 |
| Mumbai | 0.62 | 0.62 | 0.64 | 0.66 | 0.66 | 0.67 | 0.67 |
| Seoul | 0.54 | 0.56 | 0.59 | 0.60 | 0.62 | 0.63 | 0.63 |
| Shanghai | 0.40 | 0.41 | 0.41 | 0.42 | 0.44 | 0.45 | 0.45 |
| Tehran | 0.47 | 0.52 | 0.55 | 0.59 | 0.61 | 0.64 | 0.65 |

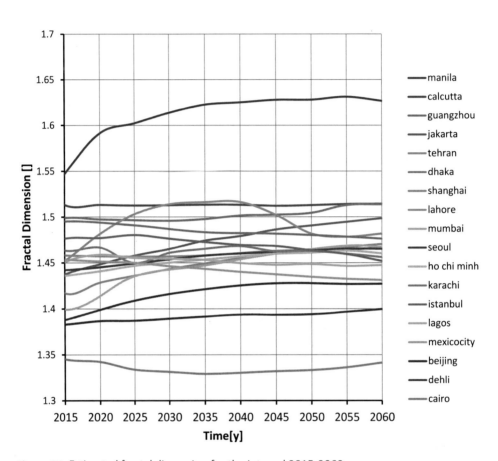

**Figure B1:** Estimated fractal dimension for the interval 2015-2060

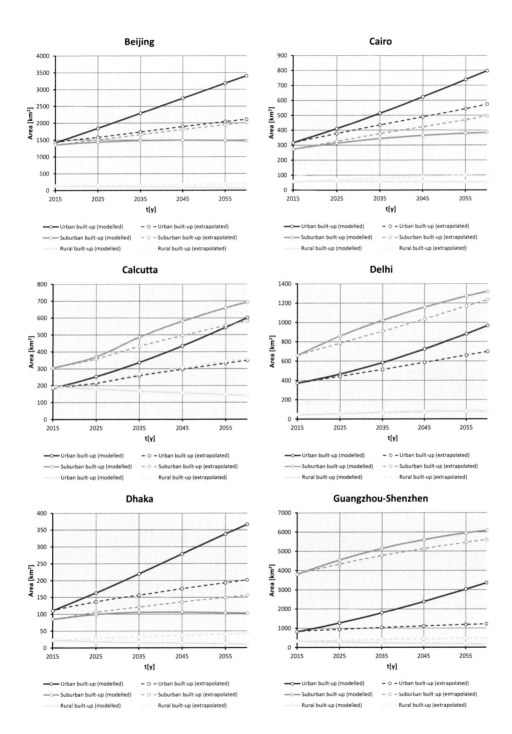

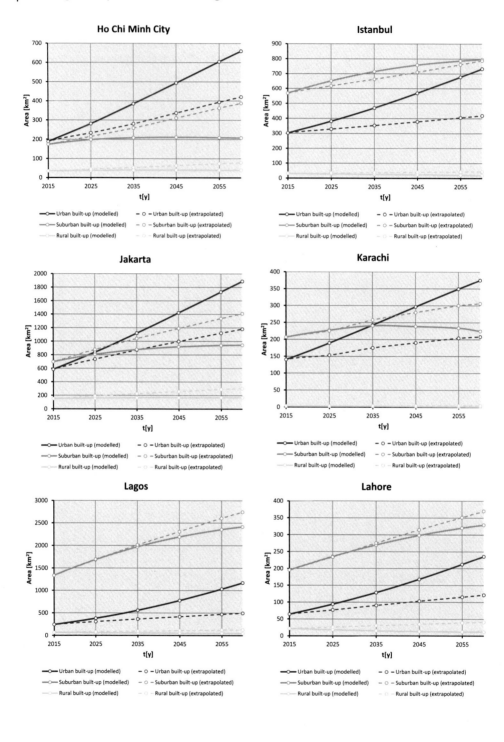

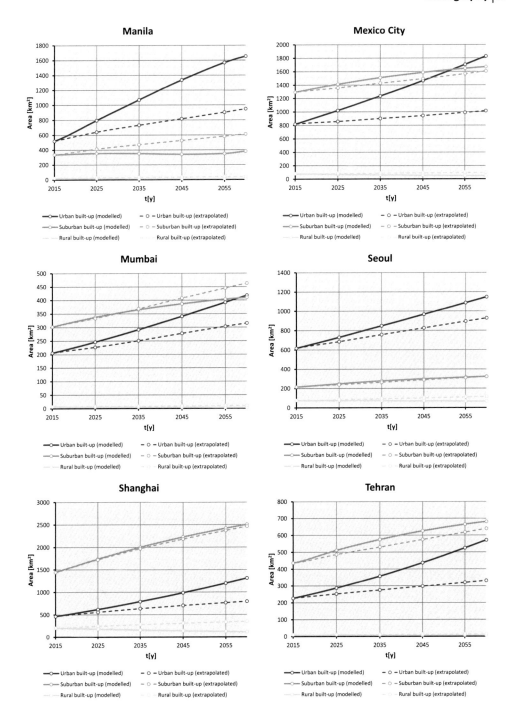

# Acknowledgements

I've always been fortunate enough to be able to consider my PhD as a hobby. This provided the freedom to choose my own topic and also to pursue projects that could explore and enrich the topic in a way that would not have been feasible if I'd been squeezed in a regular full-time PhD position. One of the projects that contributed most to my research was the FP7-project CORFU, where under the guidance of the wonderfully cool Slobodan Djordjević, I took my first steps in urban modelling in Beijing, Dhaka and Mumbai. From there on, conquering the rest of the world was only a small step.

Yet, that conquest would have led me knowhere if it wasn't for the three wise men that taught me how to maintain an insatiable appetite for knowledge (and many other things), be ever flexible in my thinking and open my eyes to areas that were dark an murky before: Richard Ashley, Erik Pasche and Chris Zevenbergen. While sadly enough Erik passed away way too soon, his drive and passion are still radiating today. Fortunately, I'm still able to enjoy massive power surges from Richard and especially Chris, who still manages to bury me under a creative avalanche of ideas and initiatives (Snap those fingers Chris...."This is IT!")

My gratitude also goes to IHE-Delft, the institute that has been instrumental in making this vast world my oyster. On those many journeys across the world, I've met ever so many inspiring people that I am fortunate enough to call my friends: Jason, Kirill, Quan, Polpat, Sachi, Taneha and many others.

A shout out also goes to my Rotterdam-based friends that always contributed to making my life better (although not necessarily my PhD): Annelore, Agnes & Stefan, Anja & Bertrand, Deborah & Michel, Giant, Nanouk & Daan, Olaf & Saar, Sieta & Coen (miss the Moscow madness), The IFFR-crew and the every hospitable Volmarijnstraat (maybe I can reinstate "de fillem van Ome Willlem" now). It obviously goes without saying that gratitude also goes to my small but warm family: Hannie & Erik and my wife Irina.

Last but not least though, I have to thank the FRG-team which with limited resources gets more things done than any group of professionals I know: Assela, Berry, Mohan, Jeroen, Kim, Pieter & Sebastiaan and Natasa (FRG-Big). I hope I can keep contributing for many years to come. "You've done a man's job sir..."

# Curriculum Vitae

Mar. 16, 1970    Born in Doetinchem as WILLEM VEERBEEK.

**Education**

2002-2007        MSc in Artificial Intelligence, VU University, Amsterdam, Netherlands (unfinished)

1998-2001        MSc Architecture (distinction). Delft University of Technology, Netherlands

**Employment Record**

2014 – present   Project Coordinator, Flood Resilience Group. Chair Group of Water Science & Engineering. UNESCO-IHE, Delft, Netherlands.

2007 – present   Associate researcher in Urban Flood Management, Flood Resilience Group. Chair Group of Water Science & Engineering. UNESCO-IHE, Delft, Netherlands.

2005 – 2014      Project Leader Urban Flood Management. Dura Vermeer Business Development. Hoofddorp, Netherlands.

2003 – 2005      Co-Founder. DIN-arch. Rotterdam, Netherlands.

1995 – 2005      Co-Founder. Vertex Architecture & Planning. Rotterdam, Netherlands.

1994 – 1997      Junior Designer. NOX Architects. Rotterdam, Netherlands

**Experience Record**

2016-2020        BEGIN: Blue Green Infrastructure through Social Innovation. Interreg VB North Sea Programme.

2016-2020        DeltaCAP Bangladesh. Capacity Development for Sustainable Delta Management and the Bangladesh Delta Plan 2100. Nuffic

2016-2021        'CORE Bangladesh: COmmunity REsilience through rapid prototyping of flood proofing technologies', Urbanising Deltas of the World 2 project, NWO

2015-2016        Cities, Water and Governance (Steden, Water & Sturing). Comprehensive comparative flood risk study on megacities, Netherlands Environmental Assessment Agency

2015-present     C2C exchange for disaster resilience, Ministry of Infrastructure and Environment in collaboration with UNISDR

| | |
|---|---|
| 2014-2017 | CRC Water Sensitive Cities, D4.1: Strengthening educational programs to foster future water sensitive cities leaders, Australia, Vietnam, Netherlands |
| 2013- 2016 | Bangladesh Delta Plan 2100, Ministry of Foreign Affairs of the Kingdom of the Netherlands & Ministry of Planning of Bangladesh |
| 2011–2017 | Communities and Institutions for Flood Resilience, Bangladesh-Netherlands, WOTRO-Integrated Programme |
| 2011–2012 | Climate Proofing Cities 2.3: Sensitivity and Vulnerability of Urban Systems, Knowledge for Climate |
| 2011–2012 | HSRR3.1: Adaptive development strategies for the unembanked areas in the hotspot region of Rotterdam, Knowledge for Climate Programme |
| 2011–2014 | Collaborative Research on Flood Resilience in Urban Areas (CORFU), Bangladesh, China, France, Germany, India, South-Korea, Taiwan, FP7 EU |
| 2010-2011 | Heat Stress Assessment Tilburg, Netherlands Environmental Assessment Agency |
| 2009-2010 | HSRR02: Flood risk assessment in unembanked areas in The Netherlands, Knowledge for Climate Programme |
| 2009-2010 | Delta cities: International partnership project setting up learning alliances between flood prone cities focusing on UFM issues, Min. Transport, Public Works and Water Management |
| 2009-2010 | Delft Spetterstad, Living with water Programme; |
| 2008-2009 | Climate Proofing The Netherlands, Dutch Environment Agency / Climate Changes Spatial Planning Programme |
| 2008-2009 | Feasibility Study 'Rode Waterparel', Zuidplaspolder, Netherlands. |
| 2006-2010 | COST C22: Urban Flood Management, EU COST-action programme. |
| 2005-2006 | Watervast, Competition for new ideas on the use of river-sediments, Dura Vermeer, Alterra, Robbert de Koning, |
| 2004-2005 | Meer dan Duin, Regional Netherlands. development strategy for the Wieringermeer (North-West of the Netherlands), Rabo real estate, AM project development |
| 2004-2005 | Structuurvisie Den Haag, Large scale planning policy for the city of The Hague, Municipality of The Hague |
| 2004-2005 | Verkeersplan Tilburg, Netherlands. Traffic plan for the city of Tilburg, Municipality of Tilburg; |

| | |
|---|---|
| 2003-2004 | Ontwikkelingsplanologie Wassenaarsche Polder, Large scale design proposal for rural district, Dura Vermeer, AM project development. |
| 2002-2003 | Masterplan Brainpark Hengelder, Didam, Netherlands. strategy for development of the extension and reorganization of a brain park, located at the A-2 corridor, municipality of Didam |
| 2001-2002 | Designlab of the Future, Rotterdam. Research on automated future housing and urban development systems, Ministry of Spatial Planning, Municipality of Rotterdam, Netherlands |
| 2001-2002 | Urban Masterplan Philips High-Tech Campus, Waalre, Netherlands. First stage planning for the transformation and extension of the Natlab Complex, Philips Real Estate |
| 2001-2002 | Anthony Fokker Business Park, Schiphol. Master plan for the transformation of the former factory Area, Delta Project Development, Schiphol, Netherlands |
| 1995-1997 | H2OXPO, Design of water exhibition centre, Neeltje Jans, Netherlands |

# Publications

## Peer-reviewed journals

Chen, A. S., Hammond, M. J., Djordjević, S. Butler, D., Khan, D. M., Veerbeek, W., (2016). From hazard to impact: flood damage assessment tools for megacities, Natural Hazards 82 (2), 857-890

Khan, D. M., Veerbeek, W., Chen, A. S., Hammond, M. J., Islam, F., Pervin, I., Djordjević, S. & Butler, D. (2015). Back to the future: Assessing the damage of 2004 Dhaka Flood in the 2050 urban environment. Journal of Flood Risk Management. In Press. DOI: 10.1111/jfr3.12220

Veerbeek, W., Pathirana , A., Ashley, R. and Zevenbergen, C. (2015) Enhancing the calibration of an urban growth model using a memetic algorithm, Computers, Environment and Urban Systems 50 (March 2015), pp53-65

Ahmed, F., Gersonius, B., Veerbeek, W., Khan, M. S. A., & Wester, P. (2015). The role of extreme events in reaching adaptation tipping points: a case study of flood risk management in Dhaka, Bangladesh. Journal of Water and Climate Change, 6(4), 729-742.

Pathirana, A., Denekew, H. B., Veerbeek, W., Zevenbergen, C., & Banda, A. T. (2014). Impact of urban growth-driven landuse change on microclimate and extreme precipitation—A sensitivity study. Atmospheric Research, 138, 59-72.

Zevenbergen, C. Van Herk, S., Rijke, J., Kabat, P., Bloemen, P., Ashley, R., Speers, A., Gersonius, B. and Veerbeek, W. (2013) Taming global flood disasters. Lessons learned from Dutch experience, Natural Hazards 64 (3), pp127-1225

Veerbeek, W. and Zevenbergen, C. (2009). Deconstructing urban flood damages: increasing expressiveness of flood damages models combining a high level of detail with a broad attribute set, Journal of Flood Risk Management, 2(1), 45-57

Zevenbergen, C., Veerbeek, W., Gersonius, B. and Van Herk, S. (2008) Challenges in Urban Flood Management: travelling across spatial and temporal scales. Journal of Flood Risk Management (2008) 1:2(81-88)

## Book contributions

Zevenbergen, C., Veerbeek, W., Gersonius, B., Thepen, J., and  van Herk, S. (2008) Adapting to climate change: using urban renewal in managing long-term flood risk. in: Flood Recovery, Innovation and Response. D. Proverbs, C.A. Brebbia and E. Penning-Rowsell (eds), WIT Transactions on Ecology and the Environment, Volume 118, ISBN: 978-1-84564-132-0.

Veerbeek W., (2007) Flood Induced Indirect Hazard Loss Estimation Models. In: Advances in Urban Flood Management. R Ashley, S Garvin, E Pasche, A Vassilopoulos, and C Zevenbergen (Eds.). Taylor and Francis. ISBN: 978 0 415 43662 5.

Veerbeek, W., Zevenbergen, C., Gersonius, B. (2010). Flood risk in unembanked areas. Part C Vulnerability assessment based on direct flood damages. CfK 022C/2010. ISBN 978-94-90070-24-3.

## Conference proceedings

Htun, K. Z., Van Cauwenbergh, N. and Veerbeek, W. (2016) Urban Landscape Dynamic Analysis on Mandalay City, Myanmar. International Conference on the Mekong, Salween and Red Rivers. 12 November 2016, Mandalay, Myanmar

Veerbeek, W., Bouwman, A. and Zevenbergen, C. (2016) The contribution of future urban growth to flood risk in megacities. he Second International Symposium on Cellular Automata Modeling for Urban and Spatial Systems (CAMUSS), 21-23 September 201, Québec, Canada.

Nilubon, P., Veerbeek, W., & Zevenbergen, C. (2016). Amphibious Architecture and Design: A Catalyst of Opportunistic Adaptation?–Case Study Bangkok. Procedia-Social and Behavioral Sciences, 216, 470-480.

Veerbeek, W., Gersonius, B., Chen, A.S. and Khan, D.M. (2014) Applied Flood Resiliency: A method for determining the recovery capacity for fast growing megacities, 13th International Conference on Urban Drainage, 7-12 September 2014, Kuchin, Sarawak, Malaysian Borneo

Veerbeek, W. and Zevenbergen, C. (2013) Developing business-as-usual scenarios for the urban growth of four Asian megacities, International Conference on Flood Resilience (ICFR), 5-7 September, Exeter, UK

Kurzbach, S, Hammond, M., Mark, O., Djordjevic, S., Butler, D., Gourbesville, P., Batica, J., Veerbeek, W., Birkholz, S., Schlitte, F., Chen, A., Manojljović, (2013) The development of socio-economic scenarios for urban flood risk management, N. NOVATECH, 8th International Conference, June 23-27, Lyon, France

Veerbeek, W., Bachin, T., Zevenbergen, C. (2012) Exploration and exploitation in urban growth models, Optimizing CA-based urban growth models using a memetic algorithm. In Pinto, N. N., Dourado, J. and Natalio, A. (eds) Proceedings of CAMUSS, International Symposium on Cellular Automata Modelling for Urban and Spatial Systems, November 8-10, Porto, Portugal

Veerbeek, W., Pathirana, A., Zevenbergen, C. (2012) Natural Hazards Impact and Urban Growth, Legitimacy and urgency for connecting urban growth models to natural hazard impact assessment. In Pinto, N. N., Dourado, J. and Natalio, A. (eds) Proceedings of CAMUSS, International Symposium on Cellular Automata Modeling for Urban and Spatial Systems, November 8-10, Porto, Portugal

Veerbeek W., Denekew H. B., Pathirana, A. Brdjanovic D., Bacchin, T. K., Zevenbergen, C. (2011). Urban Growth Modelling to Predict the Changes in the Urban Microclimate and Urban Water Cycle. 12th International Conference on Urban Drainage, September 11-16, Porto Alegre, Brazil

Veerbeek, W., Zevenbergen C. and Herk van S. (2007): Urban Flood Management: Towards a Flood Resilient Urban Environment. In: SIBICO International (eds) Conference Proceedings Water resources systems management under extreme conditions, Moscow, Russia,pp 465-471

### Technical reports

Veerbeek, W., Husson, H. (2013) Vulnerability to climate change: appraisal of a vulnerability assessment method in a policy context. KfC report number: 98/2013 (Unesco-IHE OR/MST/177). Knowledge for Climate Programme Office, Utrecht.

Ven, F. van de, Nieuwkerk, E. van, Stone, K., Veerbeek, W., Rijke, J., Herk, S. van, Zevenbergen, C. (2011) Building the Netherlands climate proof: urban areas. KvK 042/2011 - KvR 036/2011. Knowledge for Climate Programme Office, Utrecht.

Stone, K., Duinen, R. van, Veerbeek, W., Dopp, S. (2011)    Sensitivity and vulnerability of urban systems : assessment of climate change impact to urban systems. Deltares, Delft.

Veerbeek, W., Zevenbergen, C., Gersonius, B. (2010) Flood risk in unembanked areas. Part C: Vulnerability assessment based on direct flood damages. KfC report number KfC 022C/2010. Knowledge for Climate Programme Office, Utrecht.

Veerbeek, W., Gersonius, B. (2010) Flood impact assessment for the Rotterdam unembanked areas. Knowledge for Climate Programme Office, Utrecht.

Veerbeek W. (2010) Determining the adaptive capacity for flood risk mitigation: a temporal approach. Knowledge for Climate Programme Office, Utrecht.

Veerbeek, W., Huizinga, J., Asselman, N., Lansen, A.J., Jonkman, S.N., Meer, R.A.E. van der, Barneveld, N. van (2010) Waterveiligheid buitendijks synthese Flood risk in unembanked areas synthesis. KvK rapportnummer KvK022/2010. Knowledge for Climate Programme Office, Utrecht

Rijke, J., Zevenbergen, C., Veerbeek, W. (2009) State of the art Klimaat in de Stad. KvK 007/09. Knowledge for Climate Programme Office, Utrecht

For Product Safety Concerns and Information please contact our EU
representative GPSR@taylorandfrancis.com Taylor & Francis Verlag GmbH,
Kaufingerstraße 24, 80331 München, Germany

Printed and bound by CPI Group (UK) Ltd, Croydon, CR0 4YY
01/05/2025
01858613-0001